# To Free the Mind

# TO FREE THE MIND

## Libraries, Technology, and Intellectual Freedom

**Eli M. Oboler**

Foreword by
Judith F. Krug

LIBRARIES UNLIMITED, INC.

Littleton, Colorado
1983

This book was edited and proofread by the publisher after the author's death.

LIBRARIES UNLIMITED, INC.
P. O. Box 263
Littleton, Colorado 80160-0263

---

**Library of Congress Cataloging in Publication Data**

Oboler, Eli M.
    To free the mind.

    Bibliography: p. 111
    Includes index.
    1. Library science--Philosophy. 2. Library science--
Technological innovations. 3. Information science--
Philosophy. 4. Freedom of information. 5. Communica-
tion--Philosophy. I. Title.
Z665.023    1983    020'.1    83-19889
ISBN 0-87287-325-0

*To Pierce Butler and Jesse Shera,*

*in memory of and homage to*

*their consistently humanistic teaching of*

*librarianship as an art and a profession.*

"The problem for man today is to use his individual consciousness of natural processes and of his own historic nature to promote his own further growth.... What will distinguish the present effort to create world culture is the richness and variety of the resources that are now open, and the multitude of people now sufficiently released from the struggle for existence to play a part in this new drama."

—Lewis Mumford, *The Transformation of Man* (1956)

# TABLE
## of
# CONTENTS

# FOREWORD

## Eli M. Oboler, 1915-1983

"An eloquent and insightful champion of intellectual freedom." "Our loyal 'gadfly.'" "A passionate defender of First Amendment principles." All of these terms were used at various times to describe Eli M. Oboler and his long, illustrious, and colorful career. For over three decades Eli was University Librarian at Idaho State University (Pocatello); his legacy to that institution was summed up by Dr. Myron Coulter, Idaho State University president, when he praised Eli for service "not only as a librarian, but as an advisor, as a sharp-witted person with a sense of humor and, more importantly, a man with a sense of purpose. He will be missed very much by the university, the community of Pocatello, and by all persons who are interested in our individual freedoms."

But Eli Oboler's impact extended far beyond Idaho State University. In truth, he was an exemplar of service to his profession: the list of his publications and of the offices he held will long remain a road map of the library profession and, particularly, its commitment to intellectual freedom. A man who constantly and impatiently demanded the dismantling of all barriers to freedom of expression and inquiry, he lives on as our untiring conscience.

Eli Oboler was educated at the University of Chicago and the Columbia University School of Library Service. He was a member of the ALA Council from 1951-1959 and again from 1977 to 1981. He served as a member of the ALA Intellectual Freedom Committee from 1965 to 1969, and as chair of the Intellectual Freedom Round Table, 1980-81. He was a past president of the Pacific Northwest Library Association and, for nearly a decade, editor of its quarterly. He also was a past president of the Idaho Library Association and, later, was its Intellectual Freedom Committee chair for many years.

A charter member of the Freedom to Read Foundation, Eli served on its board of trustees from 1971 to 1975 and again from 1976 to 1980. He was the only person ever elected to the board for two consecutive two-year terms on two different occasions. Culminating his service on behalf of the foundation, Eli served as vice-president, 1979-80.

The bibliography of Eli Oboler's publications is too lengthy to enumerate fully, for during his career he contributed to nearly every significant professional publication. Eli, himself, once estimated that he had published as many as five hundred book reviews alone. He was a frequent contributor to the *Newsletter on* xii–TO FREE THE MIND

*Intellectual Freedom*, published by the ALA Intellectual Freedom Committee, as well as to *Library Journal*. A selection of his writings for the *Newsletter* and other publications, *Defending Intellectual Freedom: The Library and the Censor*, was published in 1980. In 1981 he edited an important compendium of materials, *Censorship in Education*. But perhaps the most enduring Oboler publication is his now classic 1974 study *The Fear of the Word: Censorship and Sex*.

In April, shortly before his death, Idaho State University honored Eli by renaming the library which he served for so many years the Eli M. Oboler Library. I was privileged with an invitation to participate in the dedication ceremony, thus having an opportunity to publicly acknowledge Eli's monumental contribution to intellectual freedom. In my remarks, I said, "Eli is a purist. This is not to say, however, that he fails to see the complexities of First Amendment issues, for his commentary has been consistently characterized by perception. Indeed, his mind is as penetrating as it is principled. But by calling him a purist, I acknowledge that Eli refuses to be swayed by the winds of pragmatism which always flow strongest at the very time when above all we are called upon to stick firmly to principles."

Upon learning of Eli's death, Freedom to Read Foundation President William D. North sought to encapsule the essence of Eli's creed: "He saw the truth as man's ultimate and never ending quest. He saw the quest for truth as the ultimate means to free the mind and soul. He saw the quest for truth as the linkage of the ages and generations and civilizations through which the best of mankind lives forever for the betterment of mankind."

Scholar, thinker, activist, and educator, Eli Oboler will be remembered wherever the freedom to read, to investigate, and to think is cherished. His voice was rarely quiet, but was always civil, penetrating, and principled. It conveyed Eli's personal integrity and his beliefs in reason and tolerance. And in the final analysis, these traits serve as "the continuing legacy of his wisdom."*

1983

Judith F. Krug
Director
Office for Intellectual Freedom
American Library Association

---

*ALA Council, *Memorial Resolution for Eli M. Oboler*, adopted June 29, 1983.

# PREFACE

A well-known ancient Chinese ideograph signifies *both* "problem" and "opportunity." What comes out of the present welter of developing library technology can be either or both. The pages that follow are a sincere effort to face the problems inherent in converting an essentially humanistic endeavor—librarianship—into a technological pursuit, one that will still be founded on the people-serving, people-based principles and practices that have dominated the modern American library's first century-plus. There is more here than a mid-20th century curmudgeon's Jeremiad on the evils of mechanization. This is a heart-felt call to librarians to remember their professional heritage and obligations, to deal with computers and telecommunications and all their related paraphernalia as just that—equipment, apparatus, tools—rather than as some kind of sacrosanct idols whose oblations include the sacrifice—whether deliberate or unconscious—of humanistic ideals. The bedrock responsibility of the librarian is to serve as many people as possible with as much information as possible for as little cost as possible. Machines and networks and miniaturization will help the librarian in achieving this aim, if their dangers, as well as their benefits, are acknowledged and dealt with intelligently. It is my hope that this little volume will be of help in that endeavor.

This book—to forestall some obvious criticism—is certainly not up-to-date. It could not be, in this field, unless it were issued in loose-leaf form, with timely supplements appearing just about every day after its publication. I have tried to select exemplary reports and comments, from the voluminous literature on the topics discussed, that would have more than a quotidian value.

Special acknowledgements should be made to those library periodical editors who have encouraged my efforts in thinking through some not-too-popular ideas by publishing various portions of this book along the way. The faculty and administration of Idaho State University, who have honored me by conferring emeritus status when I retired and who have helped my auctorial endeavors with a most useful office, deserve my appreciation; and, of course, a special tip of my hat to Chronos—that old Greek deity who managed to get me to an appropriate age for retirement from active administration, so that I had the time to write this book just when—I believe—it needed writing.

E.M.O.

# INTRODUCTION
## Printing and the
## Coming of Compunications

The basic question in dealing with the relationship of man and any machine is this: who controls whom? Does man govern the machine, or does the machine govern man? In terms of freedom and the mind, will the use of any machine free or enslave the mind? The mind of man, untrammeled by the multitudinous details that limit the wide horizons toward which we aspire, can achieve undreamed-of goals. Even more vital for the cause of freedom, an eternal inhibitor—the censor—finds some trouble in dealing with the mechanical side of communications.

## THE PRINTING PRESS

Before the typographical age came to Western civilization in the mid-fifteenth century, the censor—usually religious, but sometimes political or social—had a comparatively easy time. There were a limited number of *scriptoria*; the medieval manuscript was not only a work of art (especially if "illuminated"), but it was also, factually speaking, the result of a great deal of painstaking work over a long period of time. The paucity of sources (all carefully supervised), the difficulty of dissemination, the very limited amount of literacy—all helped the censor to keep the medieval mind in chains. But Gutenberg's phenomenal breakthrough, coincident with the greater availability of paper and ink, soon changed that situation. The printing press spread all over Europe, and it became a real struggle for the censor to stamp out widespread brushfires of the greatest enemy of censorship—the ready availability of knowledge. Scholars have shown that Venetian printers alone produced over two million volumes before the end of the century when European printing began.[1]

The printing press opened closed doors, unlocked chains, helped free the mind of Europe for the late Renaissance, the Reformation, and the Enlightenment. As Will Durant put it: "printing replaced esoteric manuscripts with inexpensive texts rapidly multiplied, in copies more exact and legible than before, and so uniform that scholars in diverse countries could work with one another by reference to specific pages of specific editions."[2] When Portuguese, Spanish, British, and Dutch world navigators wanted to learn more about the world that some of them

1

circumnavigated, they did not have to rely on apocryphal maps filled with "here be monsters" and *terra incognita* warnings. Instead, they had easy availability of precise maps and charts from authentic sources, things never before so readily, even conveniently, at hand.

The invention of printing, of course, had even more far-reaching results than these. The top current scholar in the field, Elizabeth Eisenstein, has detailed some of the change-producing effects including:

1) the spread of literacy.

2) an "avalanche" of "how-to" books, with hardly any parallel in pre-printing times.

3) a widening of scholarship.

4) a collaborative approach to the collection of data.

5) the ability to improve and correct works, once published, hardly ever used in scribal culture.

6) the ability to preserve knowledge against such vicissitudes as war, fire, or flood.

7) making available to all previously hidden, even secret, data.[3]

About this one, Eisenstein commented, "the notion that valuable data could be preserved best by being made public, rather than by being kept secret, ran counter to tradition, led to clashes with new censors, and was central to both early modern science and to enlightened thought."

8) the prevalence of print "arrested linguistic drift, enriched as well as standardized vernaculars, and paved the way for the more deliberate purification and codification of all major European languages."[4]

9) it promoted the trend to democracy by providing "wide distribution of identical bits of information [which] provided an impersonal link between people who were unknown to each other."[5]

10) introduced "new forms of intellectual property rights."[6]

11) stimulated self-improvement, artisanship, and invention.

12) caused more interchange between intellectual disciplines, formerly compartmentalized simply for lack of easy communication.

These are only a few of a seemingly endless series of direct and indirect results of Gutenberg's venture into perfecting the use of wood blocks to make multiple copies of indulgences and playing cards. Perhaps the most truly significant basic change in the Western world was the shift from faith to reason, from acceptance of whatever was promulgated as truth by authority, to the scientific approach, which required experimentation and research for wide acceptance.

As Eisenstein has put it, "the new print technology ... made food for thought more abundant and allowed mental energies to be more efficiently used."[7] Since the advent of print, there has been available "an open-ended investigating process, where increments of new information accumulate and the frontiers of the unknown recede...."[8] Modern science itself, all authorities agree, would have been impossible without printing. It was no longer the individual—whether Pope or commoner—who decided what existed in nature or technology; the collective free mind of science reported observations and measurements. These were so widely available that all could accept or dispute, and none had to accept blindly pronunciamentos from a single, supposedly omniscient source. Were it not for printing, we might still accept Aristotle's dictum that worms come out of horsehairs placed in water, or pre-Galilean versions of our cosmos. True, the early use of printing was mainly to make available old texts, but it soon blossomed forth with what Eisenstein describes as "fresh records and drawings made by careful observers in an easily portable form."[9]

All this is not to say, of course, that the rise of printing meant the decline of censorship. Indeed, within less than a hundred years after Gutenberg's first printing venture, both church and lay authorities all over Europe practiced book control.[10] The city of Mainz (birthplace of the printing arts) was also (together with Frankfurt) the site of the first (1486) secular censorship office. It is the first known time in the post-Gutenberg era that censorship was used as a prerogative of the state. The very first edict against *printed* books (emanating from Frankfurt) was "aimed at the suppression of Bible translations into the vernacular...."[11]

Of course, the ecclesiastical authorities simply continued their ongoing efforts to suppress "heretical and schismatical publications."[12] In 1543, Cardinal Caraffa, leader of the recently revived Inquisition, issued a decree that "henceforth no book, old or new, regardless of its content, should be printed or sold without the permission of the Inquisition."[13] And the first *Index Librorum Prohibitorum* was issued by Pope Paul V (formerly Cardinal Caraffa) in 1559. This is not the place for a blow-by-blow account of what censorship has done to the freedom of the mind through the ages.[14] Suffice it to say that throughout the already over half-millenium of the history of printing, the efforts of censors to deny access to what they were against have always met (and, it is to be hoped, will always meet) an equal and often superior reaction from promulgators of opposite points of view.

S. H. Steinberg tells an amusing anecdote about just such a confrontation. The censors in Karlsbad, Germany, in 1819, tried to stifle the liberal movement in Germany by imposing a preventive censorship upon all political pamphlets of less than 320 pages. Immediately the publishers and printers of Germany "rather successfully counteracted this measure by using the smallest possible size ... [and] the largest possible types...."[15] This is only one more instance of man's eternal struggle for free expression, one in which formal, inhibiting rules ran into flexible minds—and the censor did not prevail.

## COMPUNICATIONS

The next great stimulus to learning—a long, long time after printing—was telecommunications, which brought with it its own congeries of good and bad results, especially as related to intellectual freedom. But there problems were intensified by the combination with what has, not yet justifiably, been called "the thinking machine," the computer.

Anthony Oettinger has coined a term that subsumes nearly all that has happened of real consequence to information conveyance in the mid-twentieth century—"compunications." This term means the technological merging of tele-communications and teleprocessing, so that "the distinction between processing and communications becomes indistinguishable."[16] This blurring or transference of function has brought about what Daniel Bell calls an "upheaval in telecommuni-cations and knowledge...."[17] The same kind of message or information that was carried by hand, horse, or sailboat in the early Renaissance, by train or steamship in the mid-nineteenth century, and by telegraph or telephone during the last century or so, can now be electronically flashed from continent to continent or into outer space and back by "compunications."

In rather similar fashion to the Darwin dictum that "ontogeny recapitulates phylogeny" (the development of an individual organism recapitulates the develop-ment of a kind or type of living organism), so the history of compunications seems likely to go through somewhat the same developmental process that tele-communications has followed and probably will follow. As the printing press helped fight illiteracy, so TV and radio have helped promote the spread of informa-tion.[18] As the printed book was first so widely available that it seemed no censor could cope with its prevalence, so broadcasting (and the computer's becoming an integral part of it) seems, in the 1980s, to be purely an instrument for free press and speech.

But countervailing pressures need to be watched closely. The 1949 Federal Communications Commission's Fairness Doctrine for broadcasting ran exactly counter to the First Amendment's guarantee of freedom of expression for the press. As Saudek says, "in the one case, the government is foresworn from punish-ing an offending printer, while in the other case, the government sets itself up as vigilant watchdog over the broadcast ward, with the punitive power of censorship in the revocation of the license to broadcast."[19] Under President Nixon (1969-1974), the 600 stations affiliated with the three commercial TV networks were nearly forced into "being indirectly accountable to the F.C.C. for every *network* program it carried, including especially all news and public-affairs programs."[20] Such a requirement would have forced pre-presentation program editing, resulting in either direct censorship or in advance scheduling of Fairness Doctrine responses to controversial network items. Fortunately, this got no further than the planning stages; the Watergate scandal took up all presidential energies in the nick of time.

But the possibility of some kind of censorship for compunications is, of course, always present. As long as there were relatively few radio sets, who cared what came out of them? As long as telecasting was not seen as politically or morally effective, the censor didn't bother. But with satellites and "narrow-casting" and the whole country on the verge of becoming "the wired nation,"[21] there is no longer any question that censorship is an imminent—if not already a current—problem.

When every home has its own access to practically everything, via what is going eventually to be a "wired world,"[22] then the censor will really have a field day. The central problem, as one editor has put it, "if not content, is likely to be control."[23] Whoever puts up and controls the satellites that make "compunication" possible on a global basis, it is evident, will control what can be seen and heard on any individual viewer's receiver. The stakes are gigantic; players will include multinational corporations, governments, and the public in general. At this time, no clear guidelines exist as to whom will dominate or regulate what has been called "the electronic communications thoroughfare." Already at least one U.S. senator has tried to put the whole broadcast/telecast/compunications industry under the mantle of the First Amendment, rather than at the tender mercies of a possibly politically motivated FCC.

## TECHNOLOGY
## AND INTELLECTUAL FREEDOM

From Gutenberg to the latest developments in transmogrifying a chip of silicon into a marvel of complex technology, an unvarying basic philosophic problem has been present, as mentioned briefly earlier in this chapter. Is the cause of intellectual freedom helped or hindered by the late twentieth-century developments on many fronts of new ways to send, receive, store, and disseminate widely the vast amounts of information now available? Will the censor find new reasons and new methods for censoring the vital communications so necessary to progress?

How technological innovation combines with old fashioned politics—the struggle for governmental power and control—is certainly significant. In a recent article, James B. Rule stated, "sponsorship ... is the key to the role of information technologies in shaping human freedom."[24] He sees four particular technologic developments as giving a clue as to who will be in charge. First, he forsees electronic payments (replacing checks) as being a likely route for possible political harassment, via the monitoring of the lives of individuals. Computerization of each person's complete financial transactions will work like this, according to Rule: "by making the collection of and access to private persons' account data cheaper and more efficient for themselves, financial institutions will also have made monitoring of individuals' lives easier for state agencies."[25] Then—shades of *1984!*—he concludes, "if precedent is any guide interests will not neglect these opportunities." Next, Rule says that "the prospect of replacing conventional mail service with a computerized electronic net should raise deep political concern."[26] When one realizes that it is not illegal today (as Rule points out) to apply so-called "mail covers" to find out who is writing to and hearing from whom, it is frightening to consider future possibilities. After all, as Rule states, "access to the 'inside' of an electronic letter would presumably be no more difficult than access to the address."

Third, Rule sees what he calls "news and home data services"[27] as offering great potential for privacy-invasion. He stresses, "there would certainly be no technological barrier to direct state monitoring of the choices of each user, much as the use of a mail or payment network might be monitored."[28] And librarians can readily see the potential of computerized lists of what people ask about and read (via the library of the future, undoubtedly electronic to some extent) in the hands of would-be or actual controllers/censors.

Finally, Rule looks at complete "electronic dossiers." They are already being used—to check welfare rolls with payrolls via the computer, for example. And, as Rule queries, "in the long run, dare we assume that these sophisticated capabilities will not be used for the harassment of political dissidents, for tracking political enemies, or identifying and discrediting opponents of the state?" He reminds us that during the Nixon/Watergate era, "one of the main thrusts of repression ... was the use of federal data files and investigatory capabilities to disrupt and harass those defined as political enemies." Of course it could happen again, and the more sophisticated the technology gets, the easier it will be to do it—and probably, in such a way as to avoid public awareness of such activities. Rule admits "that nothing about these technologies *in themselves* marks them for repressive use."[29] But, as he develops the equation, "under repressive political climates—which recur with regularity in American life—the uses of personal data themselves become repressive."[30]

Rule is just Luddite enough to question the basic need for the new information technologies—at least in use in the ways described above—if "the accumulation of such records represents a blank check to the political intentions of those who can control the systems." Where George Orwell, Rule says, "foresaw ... a world in which ruthless political interests mobilized intrusive technologies for totalitarian ends," what is happening today, says Rule, is "the development of the intrusive technologies ... on its own, without the spur of totalitarian intent."[31] Whichever way it happens, we had better be aware of the political, as well as the social, economic, and technological significance of the communication monsters we are creating ... although I would not agree with Rule on throwing out the baby with the bath water.

Rule's evidence and conclusions are disputed in a discussion by Langdon Winner. Winner warns us against what he calls "naive technological determinism—the idea that technology develops as the sole result of an internal dynamic, and then, unmediated by any other influence, molds society to fit its pattern."[32] This, of course, in simpler terms is the widely accepted truism that one can bring light and/or heat with the same match, which can start a disastrous conflagration—it's all in the intent. Winner is not quite this unsophisticated. He offers, specifically, "outlines and illustrations of two ways in which artifacts can contain political properties."[33] Without giving all the details of the illustrative technologies he tells about—including the mechanical tomato harvester-and-sorter, the 200 nine-foot-high bridges on Long Island intended to bar public busing, and McCormick pneumatic molding machines (used to break a nineteenth-century union)—let us look at his general conclusions.

According to Winner, "the things we call 'technologies' are ways of building order in our world."[34] They all have various inherent possibilities; rarely indeed do their creators, their disseminators, really think of social or political consequences. As he puts it, "because choices tend to become strongly fixed in material equipment, economic investment, and social habit, the original flexibility vanishes for all practical purposes once the initial commitments are made."[35] One has only to think about the resistance to putting into place additional national television networks to realize how true this theory is.

But Winner goes further, on what he calls "inherently political technologies."[36] It is all very well to see bad intentions in parkway plans or job-costing machinery; what about, for example, the atom bomb? Winner sees this as beyond any doubt "an inherently political artifact."[37] He cannot conceive of its existence and potential governmental use without "very rigid relationships of authority."[38] The briefcase that is always near an American president to permit him to push "the" button well illustrates this fact. Winner admits this is a very special case, but other almost equally rigid requirements affect the politics related to particular technological development. He cites the hazards and risks of nuclear power as including, potentially, "the sacrifice of civil liberties." In order to avoid the possibility of making plutonium readily available to terrorists, he warns, "workers in the nuclear industry as well as ordinary citizens outside could well become subject to background security checks, covert surveillance, wiretapping, informers, and even emergency measures under martial law—all justified by the need to safeguard plutonium."[39]

So, concludes Winner, "artifacts can have political qualities."[40] He admits that "within a particular complex of technology—a system of communication or transportation, for example—some aspects may be flexible in their possibilities for society, while other aspects may be (for better or worse) completely intractable."[41] Taking on technological innovation *without* regard for likely political (or civil libertarian) possibilities would seem like a rather foolhardy way to proceed.

## NOTES

[1] George Sarton, *The Appreciation of Ancient and Medieval Science during the Renaissance (1450-1600)* (Philadelphia: University of Pennsylvania Press, 1955), p. 4.

[2] Will Durant, *The Reformation* (New York: Simon and Schuster, 1957), p. 160.

[3] Elizabeth Eisenstein, *The Printing Press as an Agent of Change*, v. 1 (New York: Cambridge University Press, 1979), p. 116.

[4] Ibid., p. 117.

[5] Ibid., p. 132.

[6] Ibid., p. 240.

[7] Eisenstein, v. 2., p. 554.

[8] Ibid., p. 464.

[9] Ibid., p. 486.

[10] The discussion on this topic is based on Saul H. Steinberg, *Five Hundred Years of Printing*, 2nd ed. rev. (Baltimore: Penguin Books, 1966), pp. 260-72.

[11] Ibid., p. 261.

[12] Ibid., p. 262.

[13] Ibid.

[14] For some of this history, see my *The Fear of the Word: Censorship and Sex* (Metuchen, NJ: Scarecrow Press, 1974) and *Defending Intellectual Freedom: The Library and the Censor* (Westport, CT: Greenwood Press, 1980).

[15] Steinberg, p. 272.

[16] Daniel Bell, "The Social Framework of the Information Society," in Michael L. Dertouzos and Joel Moses, eds., *The Computer Age: A Twenty-Year View* (Cambridge, MA: MIT Press, 1979), p. 176.

[17] Ibid., p. 195.

[18] Robert Saudek, "TV's Future Is Reflected in Print's History," *New York Times* (Aug. 24, 1980): 33-34D.

[19] Ibid., p. 34.

[20] Ibid.

[21] Ralph Lee Smith, *The Wired Nation; Cable TV: The Electronic Communications Highway* (New York: Harper & Row, 1972).

[22] James Martin, *Telematic Society: A Challenge for Tomorrow* (Englewood Cliffs, NJ: Prentice-Hall, 1981), p. 211.

[23] Joseph Newman, ed., *Wiring the World: The Explosion in Communications* (Washington, DC: U.S. News & World Report, 1971), p. 175.

[24] James B. Rule, "The Future of Freedom: Politics and Technology," *Dissent* (Spring 1982): 202.

[25] Ibid., p. 203.

[26] Ibid.

[27] Ibid., p. 204.

[28] Ibid.

[29] Ibid., p. 205.

[30] Ibid., p. 206.

[31] Ibid.

[32] Langdon Winner, "Do Artifacts Have Politics?," *Daedalus* (Winter 1980): 122.

[33] Ibid., p. 123.

[34] Ibid., p. 127.

[35] Ibid., p. 128.

[36] Ibid.

[37] Ibid., p. 131.

[38] Ibid.

[39] Ibid., p. 134.

[40] Ibid.

[41] Ibid., p. 135.

# 1

# MICROGRAPHICS
# and
# MACROKNOWLEDGE

## The Miniaturization Miracle

One of the most unusual developments in information storage and retrieval has been microphotography and its applications, particularly for library and archival applications on microforms. Microphotography in this country came about in 1864, when John Morrow set up the first microreproduction laboratory, which was capable of producing about one microphotograph a second—even today, a pretty fast rate. The first microfiche in its current format was produced in 1906. The concept of micropublishing originated in 1853, when the British scientist, Sir John Herschel, heard of the development of microphotography and suggested "the *publication* of concentrated microscopic editions of works of reference—maps, atlases, logarithmic tables, or the concentration for pocket use of private notes and manuscripts ... and innumerable other similar applications."[1]

## DEVELOPMENT OF MICROGRAPHICS

The beginnings of commercial microfilming came in the mid-thirties, when true micropublishing began with Recordak Corporation—a subsidiary of Eastman Kodak—which with the New York Public Library put out a five-year collection of the *New York Times* on microfilm, covering the period of World War I. In 1939, the *New York Times* was offered on a regular basis on microfilm, especially to libraries. This began the microfilming of a great many newspapers here and abroad, which had several results. In the first place, newspapers generally just about stopped being bound. Secondly, all over the world, particularly in Europe, national centers were established to assemble and micropublish complete retrospective files of national major newspapers. Only in the last fifteen or twenty years has there been a really massive development of new hardware, new films, or other forms with a really concentrated attention by libraries, commercial businesses, and to a very large extent, governments, on the potential uses of microfilm.

Before 1960, library micrographics were devoted almost exclusively to the preservation, storage, and publication of "big projects," as well as the provision of single copies for those individual researchers who asked for them. Because of the development of automatic copying and the reader/printer, both librarians and library patrons began to see and use microforms as a really convenient and comparatively inexpensive way to obtain hard copies. The first breakthrough was the Xerox® copying machine, which provided hard copies of entire books and manuscripts at reasonable prices. The reader/printer gave similar service, but without the durability of xerographic copies. These new hardware and system developments in the 1960s made it possible for scholars to acquire personal collections of microforms (and their own viewers), which was certainly a breakthrough, because scholars generally had been devoted to the book. Now came microphotography and microreproduction, and things changed a good deal.

As for the future of microform, there are likely to be large, fully automated micrographic systems only in extremely large libraries. Even those libraries that have comparatively large holdings of microforms find that only a small percentage is used, and the great expense of full automation is hardly necessary.[2] There are some pending developments for automatically loading magnetic tape on computers, which might easily be adapted to film. The use of TV-type viewers ("soft display") is interdicted because they do not have the fine-grain resolution that most library microform materials require. Enlargement printers that produce prints directly from film or from a facsimile transmission again do not give very good image quality and are quite expensive. There will be use of facsimile service by libraries, but only when the office of the future shows widespread similar use by businesses, which will both improve quality and reduce costs. As the cost of building gets ever higher, space will become much more of an item to be considered, and the use of microforms will become even more widespread. There will always be a need for economizing, and there is no question that microforms contribute to this common and reasonable aim.

For many years, there has been talk of the possible $25 portable reader, both for microforms and microfiche. So far, $100 is about as low in price as any reasonably good-quality reader seems to have gotten, but there are many possibilities for development of so-called "very compact readers." The use of fiber optics, multiple lenses, ambient light—all these may result in the compact reader, but may in addition, cause greater rather than lesser expense. In summary, the real limiting factor on the use of microforms is the industrial development and manufacture of a wide variety of reasonably priced equipment. Materials are constantly being developed that give increasingly better service for microform-users, but the likelihood is that microforms will become the answer to many of the dreams of librarians only when the consumer market for entertainment, educational, and household uses of microforms becomes a reality.

Now, let's look into our crystal ball at some of the more likely possibilities for the development of micrographics. The term micrographics is generally used to cover technology that covers all aspects of compacting or compressing graphic information so that it may be recorded, stored, retrieved, and displayed for human use. The basic micrographic storage media are, of course, photographing materials. Microforms are really not substitutes for printed matter; they are certainly not substitutes for hard copy, but they do have their own values.

Originally microimages were developed as a way to preserve those works that were printed on rapidly deteriorating or vanishing paper, as well as a relatively inexpensive means to disseminate facsimile copies of documents for scholars. It was not until 1938, under the direction of Vannevar Bush, that the device called the rapid-selector was invented to search rolls of coded microfilm. Until this time, microfilm was, frankly, a pain in the neck to use. In my personal judgment, they still are not as easy to use in many ways as books, but they do have their indisputable advantages.

When we link micrographics with computers, microforms will be at their highest level of use. With the development of computer-output microfilm (COM), the microfilm has come into its own as a substitute for the card catalog. Some of these COM devices can draw illustrations, can create halftone pictures, indeed can generate almost any character style or format. Thus, microform editions of publications can be very economically created, duplicated, and disseminated, so that bringing them up to date is possible far more frequently and economically than with hard copy, whether cards, books, or periodicals. It is predicted that the impact of some current micropublishing activities will only become evident to libraries within the next few years, because more and more specialized types of publications are only available in this form. Perhaps the outstanding example is the output of the United States Government Printing Office, which is already more than 50 percent in microform and the GPO is likely to almost dispense with hard copy altogether within the next few years.

There are some continuing problems with the use of microforms, despite their obvious economy, efficiency, and ease of use. Paula Dranov, basing her opinions on a telephone and mail survey of all types of libraries throughout the U.S., offers "five major conclusions about microform usage in libraries today," as the following summary details:

1) Even though the availability of all types of information via microfilm has grown, there is no apparent nationwide trend toward library purchases of original publications on microform.

2) Of the three microforms now available, only microfiche is an evident competitor of 35mm roll film. Microcard is just about obsolescent, and "ultrafiche" (high-reduction) is not very popular.

3) Until more all-in-one reliable reader/printers that can use various types of microforms come on the market, the general use of microforms will be inhibited. Another drawback to microform use is that a need continues for more acceptable guides to proper evaluation, nationwide maintenance centers, and wider education of users on how best to use microform reading equipment.

4) The continuing "uncertainty over the suitability of non-silver films, vesicular or diazo" has prevented libraries from wide use of microforms. Although non-silver films are less expensive, their longtime durability is in question.

5) Computer-output microfilm (COM) is the most likely new general use for microforms in libraries. Further use of microforms in various library administrative and record-keeping ways is developing.[3]

# MICROFICHE, COMPUTERS, AND LIBRARIES

The combination of computer access and microfiche storage, in some situations, can provide information better and faster than any other form of knowledge access and dissemination yet known. A good example of this is the microfiche library of earth, moon, and planet photographs established about a decade ago by the California Institute of Technology's Space Photography Laboratory.[4] With literally millions of photographs available from the various projects of the National Aeronautics and Space Administration, it was considered essential to be able to provide rapid, organized access to the unique data, which could not otherwise be provided. The interactive mode of the program permits users to tell "what characteristics he wishes the pictures to have, select(s) the pictures that satisfy his criteria, and then allows him to control the display of the selected pictures and to obtain any additional information he may need to interpret the pictures."[5] (For old-fashioned book lovers, it is interesting—perhaps amusing—to note that looking at an illustrated book's table of contents and index, and then selecting, controlling, displaying, and obtaining any additional information is exactly what a codex book will permit. Of course, the *range* of selection is very much narrower.)

The CAML (Computer Accessed Microfiche Library) combines two types of technology and has been used as a prototype for similar developments. As Zimmerman says in his descriptive article, "CAML has been designed to integrate, in an easy-to-use system, the storage capacity and capability of a special microfiche viewer with the manipulating ability and speed of a computer."[6] It has achieved its purpose.

But there are difficulties with the use of microfiche. It is expensive to convert normal printed items into microfiche. As has been said, "it would be prohibitively expensive for any single library to place all its holdings on fiches."[7] And, of course, there is always the copyright problem. Converting the Library of Congress or other major collections to microfiche/microfilm would certainly involve major copyright violations.

Paul Starr points out that inevitable technological developments might solve this problem. He says, "microfiche reproduction will only deepen the erosion that photocopying began ... [for] fiche-to-fiche duplication can, in a few seconds, reproduce entire books for the price of a cup of coffee." It is the "software" that will/may do the trick. Starr predicts, "if portable readers become fairly widespread, it is highly unlikely that copyright in books, as we know it today, could be preserved." He wrote this in 1974; the advent of the videodisc provides another readily available opportunity for the kind of "vast infringement of copyright" to which he refers.[8]

The financial trade-off of microfiche versus print is amazing. Let alone the usually cited economies in storage space, cataloging, and bookmarking and maintenance, it is salutary to consider the relative costs of using books and microfiche. As far back as 1970, according to Starr, "a survey of major university libraries ... showed that the ratio of total library expenditures to the volume of general and reserve circulation indicated a cost of about $4 per book circulated.... The average book of 250 pages could be duplicated in fiche for 20 cents."[9]

Despite long-established prejudices—both by librarians and by library patrons—in favor of the book, Starr's conclusion seems wise: "The challenge at this point is to anticipate the long-run imperatives and use them as an opportunity to simplify the organization and transmission of knowledge,"[10] a dictum which could be extended to cover most of the new library technology discussed herein.

## NOTES

[1] For general discussion, see: Allen B. Veaner, "Microfilm and the Library: A Retrospective," *Drexel Library Quarterly* (Oct. 1975): 1-5.

[2] Thomas C. Bogg, "A Technological Review: The Future of Microimagery in the Library," *Drexel Library Quarterly* (Oct. 1975): 66-74.

[3] Paula Dranov, *Microfilm: The Librarians' View, 1976-77* (White Plains, NY: Knowledge Industry Publications, 1976), p. 2.

[4] R. G. J. Zimmerman, "A Computer-Accessed Microfiche Library," *Journal of Library Automation* (Dec. 1974): 290-306.

[5] Ibid., p. 291.

[6] Ibid., p. 300.

[7] Paul Starr, "Transforming the Libraries: From Paper to Microfiche," *Change* (Nov. 1974): 39.

[8] Ibid.

[9] Ibid., p. 37.

[10] Ibid., p. 40.

# 2

# NETWORKING,
# INFORMATION UTILITIES,
# and
# LIBRARIES

In 1970, a conference on information networks and interlibrary communications held at Airlie House, Warrenton, Virginia, was co-sponsored by the American Library Association and the U.S. Office of Education's Bureau of Libraries and Educational Technology. The report of its proceedings, as edited by Joseph Becker,[1] contained several suggestions and comments that have been of importance to the national plans on library/information networks ever since.

Writing on federal telecommunications policy and library/information networks, Kenneth Cox, who had been a member of the Federal Communications Commission, stressed the importance of the FCC's providing what he called "an 'open sky' for innovation...."[2] He cited a White House memorandum, sent to the chairman of the FCC on January 23, 1970, which "emphasized the need for reliable communications services for public, business, and government use at reasonable rates and the assurance of a healthy environment for continuing innovations in services and technology."[3] Here are the objectives of a viable public policy in this area, as stated by the Nixon Administration:

> Assuring full and timely benefit to the public of the economic and service potential of satellite technology;

> Insuring maximum learning about the possibilities for satellite services;

> Minimizing unnecessary regulatory and administrative impediments to technological and market development by the private sector;

> Encouraging more vigorous innovation and flexibility within the communications industry to meet a constantly changing spectrum of public and private communications requirements at reasonable rates;

> Discouraging anticompetitive practices—such as discriminatory pricing or interconnection practice and cross-subsidization between public monopoly and private service offerings—that inhibit the growth of a healthy structure in communications and related industries;

> Assuring that national security needs and emergency preparedness needs are met.[4]

This is quite an imposing and, one might have hoped, likely to be fruitful list. But the fruits were few, if any. Neither the White House nor the FCC (nor ALA) could, in 1970, have anticipated the federal anti-trust action on the Bell System, which has resulted in what (in 1983) looks like an opportunity—even a definite need—for a completely new look at the telecommunications/networking situation.

## NETWORKS AND NECESSITY

What Cox said in 1970 perhaps is of more permanent value, that interlibrary connections and other information networks are essential "if the public is to have fair and equal access to our great reservoirs of information." And the only way to achieve this is, as he said, to "develop techniques of cooperation among the various institutions involved and ... to call upon the communications industries for the network circuitry—whether terrestial or satellite, radio or cable—which (will be needed) ... to interconnect all the parts of a truly national information system."[5]

In the same report, a group led by John Bystrom of the University of Hawaii, produced the very challenging "Working Group Summary on Network Needs and Development." The most significant section in this report (at least, so far as intellectual freedom/access to information is concerned) reads as follows:

> There are staggering inequities in the sources of information that are available to people in sparsely populated, economically depressed, or educationally deprived communities, as compared with people in well-to-do, middle- and upper-class, metropolitan communities. This is true of teachers, doctors, lawyers, and engineers, as well as the lay public. *For many segments of society, there is no information agency close at hand; the available agencies are weak, or their potential users are unaware of their existence* [my italics].[6]

A little further on in the same report, they said "the United States should work toward a time when the availability of knowledge will not be severely limited by the ability of certain individuals to pay for it, by the fortunes of geographical location, or by associational membership."[7] Exactly the same comment could have been uttered when America's first beginnings of a national library movement crystallized, over a century ago. There is little question that the widespread availability of knowledge is still limited to an accidental or monied elite of information-available users.

In line with this urgent necessity, the working group recommended that "the library and information science community should promote free or low-cost telecommunication rates for educational purposes as a dividend for taxpayer investment in the development of communications technology."[8] Curiously, this as yet unfulfilled suggestion parallels the "dividend" that America's railroads—which received billions of dollars worth of free land—agreed to "pay" by granting members of the armed services lower rates for a long time after the transcontinental railroads were established. To give free telecommunication services for educational purposes seems like an equitable, good idea—but, so far, no "dividend" of this sort has been reported.

Another "working group" at the Airlie Conference provided similar recommendations bearing on intellectual freedom. They agreed, "that the seeker of information in the United States today has a right to the information he seeks unless there are legal or proprietary restrictions on the use of this information. If he is not at present served, this fact should be recognized so that he can be served. If he himself is not aware of his needs, the aim of a network should be to help him become so aware."[9]

Another interesting recommendation bearing on intellectual freedom is that "considerable care must be exercised in charging for ... services on the basis of direct cost, because serious inequities may arise if the distribution of information services is made not according to need but only according to the ability to pay."[10] At least an implication of the necessity for some kind of governmental grants to achieve this is clear in this recommendation. After all, if various federal grants through the last thirty years or so have been provided to attempt some kind of rough balancing of nationwide library access to books and other library materials and equipment, why couldn't the same theory allow for networking connection equipment and service costs? One brief statement in the conclusions and recommendations of this particular working group was about as clearly, simply, and concisely expressed as such an idea could be presented: "network services should provide access to the universe of information for all types of users."[11] Selah!

At the same conference, Edwin B. Parker, speaking of potential relationships between library and other mass media systems," stressed that libraries—and, of course, networks that connect libraries—must be able to provide *both* retrieval and storage facilities. He said that "an appropriate acronym for systems that emphasize storage without adequate retrieval would be SNARL (Storage Now and Retrieval Later)."[12] He also gave a good deal of evidence that "there will be increased need to back up local libraries if the information needs of library clients are to be met."[13]

His ingenious suggestion to take care of the "information-poor" is that "it may be desirable to provide public education subsidies to the consumer in the form of 'information stamps' that can be spent at the library or other educational institutions."[14] Parker feels that "this way favorable distribution effects can be maintained while gaining the advantages of letting people with the financial resources pay for additional service and providing an economic feedback to ensure that the library is responsive to the information needs of its clients."[15] Somehow, it boggles the mind to picture some future scandal where, presumably, one individual secured additional "information stamps" on the black market so he could get added knowledge!

## THE NATIONAL COMMISSION REPORT

After a two-year study, in 1975, the National Commission on Libraries and Information Science produced a report that they called *Toward a National Program for Library and Information Services: Goals for Action.*[16] Its basic goal was "to increase each person's access to the nation's rich knowledge resources." This was certainly a fine, egalitarian (note that they said *"each"* person) pronouncement; but, what has happened in the more than half a decade since?

To begin, the Commission's report was not blissfully optimistic. Granting that not one American library is really capable of operating without the help of others, the Commission was realistic enough to face up to a whole list of what were called "major problems," which included:

- the increased cost of acquiring library materials and organizing them for use;

- the difficulty of recruiting and compensating skilled personnel for these tasks, especially when the range of languages, subjects, and services is great;

- the growth of knowledge, with the consequent demands, particularly on academic libraries, for a wide range of specialized materials;

- the varying levels of resources and funding abilities in each state;

- the cost of storing infrequently-used materials that accumulate when a library tries to be self-sufficient;

- and the requirement to serve constituencies that are not now being served.[17]

These make a formidable, almost forbidding, group of obstacles to be overcome. But they are by no means all.

The Commission listed quite a group of "barriers and impediments which will have to be overcome to achieve the increased cooperation required to implement a nationwide network...." They began by recognizing that for effective networking the *entire* information community—going far beyond libraries—would have to be involved and cooperating. Funding for libraries and related information activities, even without what would have to be a sizable amount for networking, "is unstable and insufficient." Even if funding were available, the jurisdictional problems of matching sources and uses of money would be great. "There is danger," they said, "that a heterogeneous group of networks will emerge which may be difficult and expensive to connect, or which may never be connected at all"[18]—which should go to show that the crystal ball isn't *always* clouded!

They realized the difficulties of integrating many types of libraries together, particularly when so many are now oriented toward *not* serving the general public. The Commission was troubled by what they called the "attitudinal" problem of changing traditional librarianship from its humanistic, personal concerns with patrons to the accepting use of computer and other electronic information systems, basically an education and re-education problem. Finally, they addressed the ineluctable necessity for national networking to achieve truly national bibliographic control, "to identify items of recorded information in all media, to provide intellectual access to each such item of information, and to standardize the processing and communication of relevant data." And, along with this, was the current and continuing ignorance of the public as to just what is available where and how—another education problem. Combining all these "major problems" and "barriers and impediments," one wonders how the Commission even dared to attempt to face up to the Sisyphean task of planning a national network. One of the major reasons it did so, it is clear, was the promise of computer technology. What they recommended was a "cooperative, time-shared, multi-institutional approach to computer usage...."[19]

With all of these negative indications, it is refreshing to report that although all systems are not necessarily "go," many are at least on the verge. For example, in 1980, the Council on Library Resources released a report of a study by Batelle-Columbus Laboratories,[20] which recommended that the four most active computer-based bibliographic utilities—the Library of Congress, OCLC, RLIN, and WLN—"develop online links using the automatic translation of requests and responses."[21] This was intended to link three library operations performed by all the utilities to some extent—shared cataloging of current monographs, interlibrary loan, and reference searching.

According to the CLR study, if dedicated leased lines (using properly programmed computers to "translate" one system language into another) were tried, literally millions of dollars would be saved for the member libraries, and millions of additional successful reference searches could be made. There was one drawback to this whole scheme, however—one that, like the evils and hope in Pandora's box, would, perhaps, be a trade-off. If all four were linked, the number of potential interlibrary loan requests would be bound to increase in astronomical numbers. This, of course, raises the whole question of cost-benefits to be derived from networking and affiliation with one or more major information utilities.

Actually, "networking," as such, is not a very new idea in librarianship. From 1876 on, the libraries of America have tried to work together, particularly on a local or regional basis. What is new is the combination of cooperation (via a leased wire, or by telephone or satellite) and computerization. This nation received its greatest impetus in the late sixties and early seventies, perhaps best summarized by Eleanor Montague, who described this as the period when "libraries witnessed the development of successful networks, the sharing of computer systems, proven on-line applications, turnkey library systems, vendor-supplied automated capabilities, minicomputer applications, the promise of microcomputer applications, and greater competition among systems."[22]

## INFORMATION UTILITIES

The information utilities—those systems that recently have shown the greatest propensity to competition, rather than cooperation—are of critical importance for libraries generally. The first, of course, was what originally was called by the very unpretentious name of "Ohio College Library Center." After the coming of machine-readable cataloging data, with the Library of Congress' MARC program in 1969, it didn't take long for OCLC to add many members—fifty-four by 1970—and by 1971, it was performing online processing for several hundred members. It was the pioneer in this particular effort. And it was also the pioneer in even considering a privately-funded national library network. But OCLC was not the only group to aim at such a high goal. The Research Libraries Group (headquartered at Yale) began with the New York Public Library and the libraries of Yale, Harvard, and Columbia working together. The RLG has expanded, principally by merger with Stanford University's BALLOTS (Bibliographic Automation of Large Library

Operations Using a Time-Sharing System) to form RLIN (Research Libraries Information Network).

Two other networks started at about this time—WLN (Washington Library Network) and UTLAS (the University of Toronto Library Automation System). WLN provided its computer processing service, as does OCLC, from a central office—WLN's is located in Olympia, Washington. Some other networks are based on book-selling firms, also—IROS, the Brodart Company's book-acquisition and book-processing system, and BNA, a similar service provided to its clients by Blackwell/North America, headquartered in Portland, Oregon. The single purpose of the library network is, presumably, the sharing of resources. No one library— be it the Library of Congress or Yale University or the Moscow Library—can hold all of the books, all of the information available in a variety of forms. The pooling of resources via a computerized library network offers promise of nearly perfect equal access to information, a widely held professional and social goal.

With OCLC alone now holding in its database the online availability of over five million titles in nearly 2,000 libraries, there is promise of an eventual, not too far off solution of a number of the problems facing the information community. But is it really a solution to the nagging problems of libraries today and in the near future? Richard DeGennaro has pointed out that "resource sharing is essential but it is not a panacea."[23] What good is it after all, DeGennaro says, to save money, time, and personnel via networking, information utilities, and other means of modern resource-sharing, if nationwide library funding continues to go down in both nominal and inflation-calculated terms. This is not to say that resource sharing, or networks and information utilities are not of present and prospective benefit to the libraries of America. But, to repeat DeGennaro's point, they are *not* a panacea.

Alice Wilcox has presented some useful criteria by which individual libraries may judge the advisability of sharing in what she calls "cooperative" resource-sharing arrangements. In abbreviated form, here are her suggestions:

1) network participation should assist the library in supporting its institutional, instructional, and/or research programs;

2) (it should) contribute to the cost effectiveness of library service;

3) the library's participation in a resource-sharing network should help define its obligations to, and formalize its service agreements with, clientele outside the institutional or jurisdictional community;

4) cooperative agreements should have a positive, not adverse, impact on the library's response to the needs of its primary users;

5) the library should be (financially) prepared to fulfill its responsibilities in cooperative arrangements;

6) the library should be aware of and make appropriate use of its cooperative resources and services;

7) the library should be conscious of the implications of the cessation of a resource-sharing program.[24]

Some of these may seem obvious, even simplistic—but they are all significant. Their combined warning would seem to be—*know what you're doing before you share resources*—especially in the high-cut, sophisticated ways now available to libraries. The recent flap within OCLC's membership over the legality of third-person sharing of information is only one of the continual, unforeseen difficulties certain to arise from the new technology. In our less automated days, these were usually simple problems of interlibrary loan arrangements, mostly of a rather plain, in-house nature but not affecting an entire region or perhaps even 2,000 libraries the country over. Another aspect of this networking/information utility picture certainly merits some attention. In the last decade or so, a large number of regional networks have developed, all but WLN as part of OCLC. These include a whole alphabet soup of acronyms—from SOLINET (Southeastern Library Network) and NELINET (New England Library Information Network) to AMIGOS (Texas, Arkansas, Oklahoma, Arizona, Kansas, New Mexico) and PALINET (Pennsylvania Area Library Network). These are rapidly becoming, as DeGennaro describes them "merely subsidiaries of OCLC for marketing and training."[25]

Most important in consideration of the future place of the library network and the information utility, as I see it, is what this style of library group configuration will mean to intellectual freedom—both broadly and narrowly considered. The basic problem is, of course, that networking costs money—a great deal of it. And someone has to pay. So access is automatically restricted in relation to the amount of money available to any particular individual library. And denial of access is, of course, one way of restricting intellectual freedom. Another, related problem (referred to in chapter 4 in greater detail) is that of privacy, especially in its relation to networking. The most useful study of this problem—a little-known or cited U.S. General Accounting Office report, which appeared in 1978—concludes that "today's advanced teleprocessing technology provides the capability to store and retrieve vast amounts of information maintained about individuals and, as such, can pose a serious threat to privacy if not properly controlled."[26]

At first glance, one might not see the appositeness of this statement to the former discussion of library networks and information utilities. But let's do a little hypothesizing and conclusion-drawing. Presume that the current trend toward networking and resultant ready access to the records of a great many libraries is coupled with a continuation of the current trend toward computerizing circulation records. Then—at some not too distant dates—an individual, or a group, or some branch of government comes into your library and wants to know what information has been sought by whom, what books or other materials have been read or viewed by whom. The tug-of-war between ready, instant, nearly universal access and the right to privacy will become a very significant one at that point. A further strong possibility is an outgrowth of a service provided readily today by non-automated libraries: the request for information about a well-known individual. (We, of course, refer to such standard sources as *Who's Who* volumes and *Current Biography* and so on.) But in the heyday of the networked, computerized America, what's to prevent the library's request for and ready access to a long list of exceedingly private information in public or private databases?

All this is why I feel the GAO report is so prescient, so important. The report, in discussing the computer security problem, offered the following perceptive questioning of the dilemma presented by a government-wide teleprocessing network: "should the Government (1) take advantage of the economies that may be possible from using multi-user teleprocessing systems rather than individual agency-owned and operated data processing systems or (2) protect the individual's right to privacy by prohibiting such networks, thus avoiding the risks considered by some to be inherent in any of today's large teleprocessing systems?"[27]

They agreed with critics that "it may appear that the Government must forgo the economics to protect the rights of the individual." As a matter of fact, at several points during the seventies, various groups—executive and Congressional—felt so strongly about this that they denied even planning funds for a so-called truly "Government-wide teleprocessing network." It was all too like Orwell's *1984*—with Big Brother looking into the individual's privacy—for the report's authors.

But the GAO (and proponents of the One Big Network) see a way to avert the dangers inherent in one big computerized database that holds in its bowels just about all there is to know about you and you and you. The GAO report states, "the dilemma would be solved and economies realized if adequate controls could be defined, established, and maintained to reasonably ensure confidentiality of data."[28] As is obvious, a number of difficult hurdles must be overcome to satisfy these criteria. "Adequate" controls? Definition, establishment, and maintenance of controls? "Reasonable" ensurance of confidentiality?

Consider these one-by-one. Controls can perhaps best be defined as those against so-called "penetrators" (outside, malicious individuals) or "insiders" (those with authorized access). Of these, the GAO report concludes that "the major threat to personal privacy stems from the misuse of personal information by individuals having unauthorized access." They see only a secondary, rather unlikely, threat from malicious penetrators, who, to quote the report "penetrate a system by using an operating system function in a way unanticipated by designers, or by exploiting some anomalous behavior of the operating system."[29]

There are several methods of penetrating a supposedly closed system that do not seem too difficult to more or less anticipate. The report suggests, in this regard, "(1) acquiring by any method a list of user identifiers and corresponding passwords or other identification and confirmatory information needed to gain access to the computer system or (2) obtaining supervisory (executive or master) control of the computer system."[30] One might suppose that there are already "built-in" controls to provide safeguards against such database invasion.

However, the GAO report points out, "contemporary computer-operating systems frequently fail to provide adequate protection for personal or sensitive information."[31] Indeed, they say, "known flaws exist in several commercially available operating systems currently in use."[32] They base this rather discriminating statement on a 1976 study by the Institute for Computer Sciences and Technology, a part of the National Bureau of Standards.[33] In addition, if one is optimistic about the future, the pessimistic view from the GAO is that "there are no comprehensive criteria for security to guide those designing and implementing operating systems for computers."[34] All this gloomy talk is only too often verified by the continuing revelations of so-called "rank amateurs" penetrating—and, in some cases, stealing a great deal of money from—supposedly "secure" computerized systems related to banks. If the banks can't provide security, who can?

The GAO report does have some positive, useful suggestions for what they call "addressing the security problem." *Don't* use a network; use what they call "hardware dedicated to a single activity's use"; in other words, use a minicomputer.

The point is, of course, as the GAO report says, "informational and operational requirements may well render such an alternative impractical in many situations."[35] To approach Prometheus' gift by saying, "put out the fire and never use it" is an unlikely probability in our technological age. Since, willy nilly, the network seems inevitable, what can be done to protect what the GAO report calls "shared communication"? They suggest the use of what they call "protected domains," "hierarchial data sharing," and other quite complex but not necessarily foolproof (or theftproof) methods.

Perhaps the most practical approach is along the lines of identification, access control, and access auditing.[36] The use of data encryption—with "an encryption device at the point of data transmission and a decryption device at the point of data reception" is practical, available, but very costly. In fact, the National Bureau of Standards recommends that the three other security safeguards—identification, access controls, and access auditing—should be tried before resorting to encryption to protect personal data.[37]

The report makes clear that no such thing as complete security, or perhaps the more appropriate term in this connection, no such thing as complete privacy is possible, just as absolutely universal access is not possible. It suggests that "the major problem to be resolved by users is the definition of the proper trade-off between (1) economies achievable through the use of modern computer/communication technology and (2) the added cost of obtaining the level of protection for personal information that is appropriate for the threat involved."[38]

I have merely skimmed the surface of what may be a treasure chest or may be a quagmire for libraries. We are in the beginning stages of what seems likely to be the most startling developments in librarianship since Dewey conjured up his first decimal and Cutter prepared his first author-identifying symbol. What's past, as the National Archives' motto says, is prologue; what the twenty-first century will bring, many of you will see.

## RECENT NETWORKING STUDIES

One of the most recent authoritative studies of a library network in situ was made by Raymond F. Vondran for the Library of Congress's National Development Office.[39] The results of his study of the National Union Catalog, to a degree, vindicate some of the previous prophecies on the merits and demerits of library networks. Since the National Union Catalog was a database before the term was commonly used (1956), it is a pioneer that deserves attention—whether as a model of what should or shouldn't be done is still not quite clear. Vondran made the study to "provide background data for the systematic development of the library bibliographic component of a nationwide data base and network."[40] In order to achieve this goal, he analyzed how the NUC operated in a normal mode in order "to determine some of the requirements for an automated nationwide union catalog."[41]

What he found out should be of interest, both for its intrinsic and extrinsic implications. His conclusions were as follows for a vast operation, which in fiscal year 1978 included over 4½ million reports sent in by U.S. and Canadian libraries:

1) A set of written guidelines must be created for the editorial sections of future centers of responsibility and centers of special authorization to serve as a guide for editors and as a document to aid reporting libraries.

2) The editorial staff of a proposed nationwide network should have the most current and open access to the policies and interpretation guidelines of the LC cataloging divisions in a usable form. This access should be broadened to eventually include all libraries that report their holdings to the data base.[42]

The second of these seems quite obvious, even perhaps only "housekeeping." But the first is quite different—at least in my view. If Vondran's recommendation is followed, at least some—perhaps great—danger exists that certain types of publications may be barred from entrance into the NUC—which would seriously affect the right and opportunity for scholars, librarians, or readers to be aware even of their existence. The key phrases here are "future centers of responsibility and centers of special authorization...." These phrases could very easily provide for a Big Brother-like relationship of NUC to the libraries of North America. I'm sure that is not Vondran's—or the Library of Congress's—intention; but the possibility is there. I, for one, would feel safer—as far as intellectual freedom is concerned—with an NUC that sets no criteria for *items*—although certainly uniformity of *style* in reporting is necessary.

Another recent "Network Planning Paper"—this one by several authors—deals with document delivery,[43] a rather more difficult topic because it affects not bibliographic data but the very immediate and basic function of libraries, the delivery of actual publications. The more imaginative and gadget-minded segment of the library profession might scoff at this, since this group already pre-supposes the so-called "paperless library." But in the world of things as they *are*, this booklet is significant.

One of the authors, James L. Woods, in his contribution entitled "Document Delivery: The Current Status and Near-Term Future,"[44] describes briefly the "dimensions" of American document delivery (interlibrary loan plus the work of non-library organizations that provide documents to other organizations), and discusses, "the status of current affairs and ... trends." He then presents a most stimulating view of the near future (one to five years) of document delivery.[45]

In the first place, Woods does *not* see any strong likelihood of more than evolutionary, rather than revolutionary, developments in document delivery before 1986. Indeed, he goes so far as to say unequivocally, "while the long-term future may bring into play widespread usage of video discs and satellites or some future technology, such widespread usage probably will not become an economical option for some years to come."[46] His list of likely major changes in document delivery within the near future (five years) includes:

• Demand for documents or document copies from off-site sources will continue to increase as more online searching is done and as the percent of requests that can be filled from local stocks decreases.

- The number of requests transmitted from borrower to lender via electronic means will increase as the bibliographic utilities' data bases become accessible to more libraries and the content of these data bases continues to be enriched. This is especially true for requests for serials. More data base distributors will offer electronic ordering. It will be possible to order documents electronically directly from data base producers, e.g., ISI or CAS, or from online access to regional or subject oriented union lists.

- The number of private sector document copy providers probably will not increase dramatically. Those that exist will be the ones that provide responsive and competitively priced services. These private sector providers will capture an increasingly larger share of the total market, especially the for-profit special library segment. As more and more public sector fill sources, i.e., libraries, charge for document delivery and, as these charges increase, the cost differential between acquiring from a public versus private sector fill source will diminish appreciably.

- The amount of for-free document delivery will diminish and the amount of for-fee document delivery will grow. Borrowing institutions or their patrons or end users will be paying more, not less, to satisfy their documents needs.

- In regard to lenders' processing of requests, little change is seen. The bibliographic utilities could assist by providing certain accounting functions, but this seems unlikely in the near term.

- As postal rates increase, courier services will become more economically attractive for local and regional delivery of both requests and copies or originals.

- There will be a gradual shift to electronic delivery of copies with the introduction of high-speed telefacsimile equipment, but the percent of the total copies delivered that are transmitted electronically will remain very low.

- Selection of express delivery (USPS Express mail, Federal Express or other air express package services) for originals and copies by private sector borrowers will increase.

- There will be a few publishers that will provide electronic delivery of full text, but the aggregate will be only a fraction of a percent of the full text that will *not* be available from some kind of electronic storage. Revenues from document delivery alone cannot support the expenses associated with full text storage, retrieval and transmission.

- The libraries and commercial fill sources present today will not be supplemented by yet-to-be created fill sources. While the Center for Research Libraries may evolve into a more active participant in the overall document delivery business, especially with its computer-to-computer link to the British Library Lending Division (BLLD), the prospects of the United States having by 1986 a National Document Delivery System (read National Periodical Center) is indeed remote.[47]

Of all of these, the most significant for the continuance or strengthening of intellectual freedom is the likely growth of for-fee document delivery and the diminishing of for-free service. Every added dollar that interlibrary loan services costs the patron is an added obstacle to the widespread distribution of library materials. The libraries with patrons with money (or, perhaps even more likely, the libraries with sufficient funds to pay the piper) will get the items they need; the "have-nots"–patrons or libraries–will suffer. Even if electronic delivery of copies comes sooner than Woods believes, the cause of intellectual freedom will not be served, because such delivery–at least at first–will be far from cheap. There is really no way of being more precise about this–but the prophecy stands.

In another contribution to the "Network Planning Paper," M.E.L. Jacob, Director for Library Planning, OCLC, Inc., focuses on technology as it relates to document delivery, rather than the administrative problems addressed by Woods. Following a detailed survey on such matters as electronic transmission, readers, videodiscs, cable TV, and microform technology, Jacob asks, quite realistically, "Do we have the technology to improve document delivery? Yes. Can libraries afford this technology? Only in a few cases. Is there sufficient material in a suitable form to use these technologies? No."[48] This would seem to indicate a more or less status quo situation in document delivery.

But the final symposiast in the "Network Planning Paper" under examination thinks otherwise. Susan H. Crooks, of Arthur D. Little, Inc., writing on the future of libraries in the next twenty years or so, goes on the basic premise that "by the year 2000 substitutes could be in place for nearly all services libraries perform today."[49] Therefore, she says, "the question is not whether libraries can beat other information providers in utilizing new technology to satisfy information needs. The question is: What users' needs can libraries uniquely meet in the year 2000?"[50] The key word here, of course, is "uniquely." Unfortunately Crooks does not really answer–or in my judgment, even try to answer–her own query. What she does write about are such matters as the probable deregulation of cable television, anticipated technical improvements in VDT's, the "rapid penetration of computers in schools," and the likelihood of "a terminal in every home."

The scenario she sets up for libraries is really not a prospect for libraries, but for institutions/companies that, she predicts, will *replace* libraries. Among these are the "Classics Reading Company"[51] (which will offer "book texts via broadcast or cable delivery"), and the "Reference Questions USA"[52] (which will be a sort of super-library-reference service, serving the country from one central site via a toll-free telephone number). She foresees two developments that are perhaps closest to today's library, the "Metropolis Public Library"[53] (which handles almost exclusively non-print items) and the "Redbrick University Library"[54] (which will concentrate mostly on the development and use of electronic databases).

Finally, she prognosticates that there will be such organizations as the "Humanities Research Institute," "as ... a loose confederation of research libraries," with "a similar institute in the fields of medicine or science and technology."[55] Among her conclusions (harking back to the earlier comment about "have-nots"), is that "information service will be driven by economic factors."[56] She does say that the library of the future "should represent a broader, more diverse, and sponsor-free selection of information sources."[57] Indeed she says, "in the late nineteenth century when the citizens most in need of upward social mobility could not afford to buy more than a few books, libraries bought books for those citizens. Strong arguments can be made that there is a parallel need in the late twentieth century, and that for at least ten years, today's equivalent is personal electronic information-related luxury goods."[58]

She concludes that "there is a continuing need for libraries as institutions which are dedicated to information provision that combines: the aim to meet individuals' information needs in a way that furthers a public, organizational, or societal good with an institutional stance which allows for—within those public good constraints—objectivity and freedom from the need to make a profit."[59]

That last phrase is most important and needs to be highlighted. Perhaps something like FREEDOM FROM THE NEED TO MAKE A PROFIT might do it. The public library, the academic library, the school library, in America—with or without gadgets—*cannot* lose sight of its raison d'etre—*service to all*. The problems in maintaining this posture—as in the past—will be great in the future. But if American librarians and libraries remember their past, work in the present, and do not necessarily buy a diminished future, they—like intellectual freedom—will survive and even prosper.

## INFORMATION UTILITIES:
## A SORT OF A POLEMIC*

It is only human for library administrators—from Podunk to Princeton—to want to keep up with the Joneses. The Joneses, these days, claim that the only way to travel is in fast company—ergo, computerize. Computerize your catalog, computerize your circulation, computerize your reference, computerize your acquisitions—in a word, computerize! Then will come the fulfillment of any eschatological hopes, any messianic visions the library administrator may have secretly nourished in his or her heart of hearts. Then will come economy, efficiency, great cost-benefit ratios, accelerated speed of document delivery—you name it; computers will do it. Best of all, you can affiliate with an information utility; the very name gives promise of some kind of especially annointed authority—a solid, respectable four-square entity that will solve all your problems instanter—and give you back a nickel in change.

---

*The material in this section first appeared in "Too Big for Their Britches: Or, How Useful Is a Utility?," *Technicalities* (January 1981): 15-16.

# THE IU AGE

The signs of the IU (Information Utility) Age are all about us. The June 1980 *Library Literature* has no less than thirty items (from a paragraph to a complete issue of periodical) on OCLC, Inc., Research Libraries Information Network (RLIN), Washington Library Network (WLN), and University of Toronto Library Automation Systems (UTLAS). We are told that the information utility will (the judgment is usually "solidly" grounded on future likelihoods) deal handily with the coming of AACR 2, will handle inflation in book and serial prices, will become less expensive, and will reach many more than the present well-under-10 percent of American libraries.

However, F. W. Lancaster's *Toward Paperless Information Systems* (1978) predicts that by 1990, "as much as 40 percent of all scientific and technical publications will be transmitted from author to user by electronic means...."[60] He believes that "as much as 82 percent of all scientific-technical information will exist in machine-readable form, and as much as 92 percent of the population of generators and users of scientific and technical information will have access to online terminals."[61] Concomitant with this kind of development, of course, must be online access to the primary library instrument developed to date for bibliographical control, the card catalog.

# FOOLS RUSH IN

If this is so, then why rush into the information-utility membership? Just maybe the Yellow Brick Road to the Emerald City of all our hopes—easy access to all information—is along the online catalog road, not the video-terminal and card-catalog-set path. The only really good reason right now for joining a bibliographic information utility is if a library has an extraordinarily large backlog of books to be cataloged, and shows no sign of making inroads into it without taking some extraordinary measures. Now that the Library of Congress is closing its card catalog and going to an online catalog, we are obviously all headed in that direction— eventually. For now, however, it might be a good idea to heed the old Russian proverb, "Tisha yedish, talsha budish," which is to say, "the slower you go, the faster you come."

Nevertheless, remember that whatever subsidiary activities may be in progress or planned, the major information utilities were planned as primarily technical services tools for shared cataloging. They all use LC MARC tapes; and the feasibility of eventual online cataloging, with LC as the national center, is clearly there. The refinements necessitated by the coming of AACR 2 may be taken care of in several ways. The uniform authority lists now being developed by LC and cooperating libraries *will*, unlike those now being used by the utilities, be completely compatible with computer exigencies. This will, of course, provide for better bibliographical control than is now possible.

The combination of a new cataloging code (online at LC) and the authority-record now in process make this, it would seem, a very good time to consider seriously whether or not to join an information utility at all. After all, it is possible to keep up with AACR 2 changes, and the LC cards never were perfect. According to a recent study, "5% of all LC cards need to be modified or reprinted at least once in the first year following their original printing, 22% will require some alteration in a 10-year period, and 42% will be modified at least once in a 30-year period."[6][2] In sum, if you're the one who has to decide on whether or not to join OCLC, WLN, RLIN, or UTLAS right now, make haste slowly. Consider the options, cost-benefit ratio likelihoods, and all the practicalities of your individual library's situation. You may be trading off library services worth a great deal more than the utilities would be worth, at this time.

# NOTES

[1] Joseph Becker, ed., *Proceedings of the Conference on Interlibrary Communications and Information Networks* (Chicago: American Library Association, 1971).

[2] Kenneth A. Cox, "Federal Telecommunications Policy and Library Information Networks," in Becker, p. 7.

[3] Ibid., p. 9.

[4] Ibid.

[5] Ibid., p. 10.

[6] John Bystrom, "Working Group Summary on Network Needs and Development," in Becker, p. 14.

[7] Ibid., p. 15.

[8] Ibid., p. 16.

[9] F. F. Leinkuhler, "Working Group Summary on Network Services," in Becker, p. 85.

[10] Ibid., p. 86.

[11] Ibid., p. 90.

[12] Edwin B. Parker, "Potential Relationships between Library and Other Mass Media Systems," in Becker, p. 188.

[13] Ibid., p. 191.

[14] Ibid., p. 192.

[15] Ibid.

[16] U.S. National Commission on Libraries and Information Science, *Toward a National Program for Library and Information Services: Goals for Action* (Washington: GPO, 1975).

[17] Ibid., p. 31. Note also the "Pittsburg Study," which indicated that 40 percent of purchased items in a large academic library were not used *at all* after purchase.

[18] Ibid., pp. 36-38.

[19] Ibid., p. 55.

[20] *Linking the Bibliographic Utilities: Benefits and Costs* (Washington, DC: Council on Library Resources, 1980).

[21] "Linking the Databases," *College and Research Libraries News* (Jan. 1981): 4.

[22] Eleanor Montague, "Automation and the Library Administrator," *Journal of Library Automation* (Dec. 1978): 317.

[23] Richard DeGennaro, "Copyright, Resource Sharing, and Hard Times: A View from the Field," *American Libraries* (Sept. 1977): 430-35.

[24] Alice Wilcox, "Resource Sharing," in Robert Wedgeworth, ed., *ALA World Encyclopedia of Library and Information Services* (Chicago: American Library Association, 1980), p. 481.

[25] Richard DeGennaro, "From Monopoly to Competition: The Changing Library Network Scene," *Library Journal* (June 1, 1979): 1215.

[26] U.S. General Accounting Office, *Challenges of Protecting Personal Information in an Expanding Federal Computer Network Environment* (Washington, DC: GAO Report to Congress No. 76-102, 1978), p. 33.

[27] Ibid., p. 10.

[28] Ibid.

[29] Ibid., p. 17.

[30] Ibid., p. 12.

[31] Ibid., p. 13.

[32] Ibid.

[33] R. P. Abbott, et al., *Security Analysis and Enhancements of Computer Operating Systems* (Washington, DC: National Bureau of Standards, 1976) Report No. NBSIR-76-1041.

[34] Ibid., p. 14.

[35] Ibid., p. 19.

[36] Ibid., p. 27.

[37] Ibid., p. 28.

[38] Ibid., p. 29.

[39] Raymond F. Vondran, *National Union Catalog Experience: Implications for Network Planning* (Washington, DC: Library of Congress, Network Development Office, 1982 [Network Planning Paper No. 6, 1980].

[40] Ibid., p. viii.

[41] Ibid.

[42] Ibid., p. 43.

[43] *Document Delivery* (Washington, DC: Library of Congress, Network Development Office, 1982).

[44] James L. Woods, "Document Delivery: The Current Status and Near-Term Future," in: *Document Delivery*, pp. 1-33.

[45] Ibid., p. 3.

[46] Ibid., p. 23.

[47] Ibid., pp. 23-25.

[48] M. E. L. Jacob, "Document Delivery Technology: A Brief State of the Art Review," in: *Document Delivery* [third set of pagings], pp. 6-7.

[49] Susan H. Crooks, "Libraries in the Year 2000," in: *Document Delivery* [fourth set of pagings], p. 2.

[50] Ibid., p. 3.

[51] Ibid., pp. 6-7.

[52] Ibid., pp. 8-10.

[53] Ibid., pp. 10-13.

[54] Ibid., pp. 14-16.

[55] Ibid., pp. 16-19.

[56] Ibid., p. 25.

[57] Ibid.

[58] Ibid., p. 26.

[59] Ibid., p. 28.

[60] F. W. Lancaster, *Toward Paperless Information Systems* (New York: Academic Press, 1978).

[61] Ibid.

[62] S. Michael Malinconico and Paul J. Fasana, *The Future of the Catalog: The Library Choices* (White Plains, NY: Knowledge Industry Publications, 1979), p. 47.

# 3 EDUCATION and INFORMATION TECHNOLOGY

## The Future of CAI

A decade ago, the Carnegie Commission on Higher Education presented a report that might have been expected to produce a whole new way of viewing academic teaching and learning in our time. This report—entitled, rather grandiosely, *The Fourth Revolution: Instructional Technology in Higher Education*—predicted that "by the year 2000 it now appears that a significant proportion of instruction in higher education on campus may be carried on through informational technology—perhaps in a range of 10 to 20 percent."[1] The report went further than such figures in dealing with "off-campus instruction at levels beyond the secondary school." Here its crystal ball showed a level of electronically-associated instruction which involved eight or more out of every ten students.

For research programs, said the report, electronic technology—a term used to include videocassettes, cable TV, and computers—would be generally in use by the 1970s. For higher education administrative programs, electronic technology would be generally introduced during the 1970s and be generally in use by 1980. It would take until 1990 before the academic library would make general use of such developments, and until 2000 before it would flourish in academic instruction. It is rather easy to look back at this reverse-Cassandra (or should it be Mr. Micawber, waiting for "something to turn up") and decry the report's over-enthusiastic optimism.

But, considering that the computer as we know it is only about a quarter of a century old, videocassettes have only a decade of existence, and cable TV is only now beginning to come from under severely restricted regulations of the Federal Communications Commission, the Carnegie Commission's prognostications look pretty good. Perhaps the greatest of inhibitions on the fulfillment of the commission's fond hope is that the group specified that if instructional technology were to develop, it would require annual federal expenditures of no less than "one percent of total national expenditures on higher education" by the federal government. In these early years of Reaganomics, the prospects for such funding seems distant and dim, if not actually non-existent. Without heavy federal participation in both research and development as well as in the nationwide distribution and popularization of technology-based programs of learning, what is the likelihood of education's being widely linked with the new technology in the foreseeable future? What is the current situation?

33

Some promising signs can be found in at least one phase of the new technology, where considerable progress has been made. In what has been variously described as the "telefuture," "television-aided instructions," and "teleeducation," there is a real prospect of achievement. A recent *Change* magazine editorial went far beyond traditional educational methods and milieux in predicting that "the next great leap forward beyond the schools is teleeducation, abetted by extraordinary capabilities to deliver education to class-size audiences at per-student costs strikingly similar to those in traditional schools and colleges."[2] If one accepts the editor's basic premise—that the basic rules underlying the interface of technology with people are that the coming of new technology "almost always precedes a complete understanding of what human uses will be made of them" and that the condition for widespread use of such technological development is that "it becomes widely affordable"—then the use of television for educational purposes is exceedingly likely to be nearing its optimum within the next twenty years, or less.

It is true that the forty-odd years of so-called "open" television in the United States has produced little, if anything, worth noting beyond what's come out of the Children's Television Workshop—at least so far as educational benefits are concerned. But surely "The Electric Company" and "Sesame Street" are not the end-all and be-all of the vast potential of the electronic screen! Anyone who has seen the various feeble attempts of the various states—mainly showing old films as a gesture toward education on TV—might lose hope for the "telefuture."

There are some promising developments, however. To begin with, the recent unleashing of cable television has made various corporations look hungrily at "education" as one more way to sell their product. And when American business and industry see an opportunity for greater profits, the fallout for social purposes may well be greater, also. And the commercial networks—particularly ABC and CBS, as of 1983—have begun to show a real interest. Several recent experiments in this area seem worth mentioning. The National Education Association in 1980 agreed to work with ABC to develop current-affairs educational videocassettes. It was expected that sometime in 1981, such cassettes, featuring well-known TV news people, would be issued forthrightly—but that schedule has not quite worked out. ABC has also begun to put the world-famous Berlitz language courses on tape; and since 1975 Time/Life Films has produced videocassettes intended for continuing education.

In 1981, the videodisc began to be commercially produced by several media-device manufacturers and distributed nationwide at comparatively low prices, for both the hard- and software involved. When the cost of a normal film presentation gets under $20, on the average, it is clearly easily available to most educational institutions—particularly when the machine necessary to use it is sold for under $500.

But the most important single element for bringing the benefits of telecommunications to school and student is the satellite. Some time ago, the Public Broadcasting Service began to use satellites for its transmission of national programs, and it now plans—funding permitting—to put together all of its strictly educational programs into a special network—PTV-3. This is all very promising.

Yet, perhaps the most likely route of all for educational programming in the next decade is from satellite to cable. The obvious advantage of such a combination is that it will combine low transmission costs for video with low costs for what educational technologists still consider to be most necessary for true education—feedback, two-way audio. When the teacher and the student—no matter how far apart—can react and interact, based on the same educational presentation, then effective learning seems likely to be at its best.

One of the basic technical problems with relying on satellites is that in the present state of satellite access, very little space is available. Nearly every one of the national cable systems—and there are beginning to be a great many—must find available time on only twenty-two transponders available in the RCA Satcom I satellites. The other satellites are devoted to military or private business interests, so they are not useable for educational or even entertainment purposes.

Back in 1961, Anthony G. Oettinger, perhaps the most convincing of those who have faith in the efficacy and early use of CAI (Computer Assisted Instruction), presented what he called "A Vision of Technology and Education." This was based on the extremely simple—yet grandiose—assumption "that anything available in any library can be made available to anyone anywhere within what he thinks is a reasonable time."[3] Granted this, a great deal can be extrapolated—both for good and evil. Oettinger makes the extremely useful point that basing an educational system on remote information storage obviously would raise the possibility of comparatively easy centralized control over what is taught and how it is taught. As he says, "one need only picture the use a Hitler or a Stalin could have made of a national education informational pool to understand the seriousness of this problem."[4] And the further question arises: would not a Hitler or a Stalin be more likely to arise more often if the extreme variety and hetereogeneity of current, traditional educational methods characteristic of democracy, turn to Big Brother-type sources? But this is all part of the difficulty with the spread of the denial of intellectual freedom that underlies all the new technology as it relates to education and libraries.

An example of computer-assisted instruction, making use of the PLATO (Programmed Logic for Automatic Teaching Operations) system, was developed at the University of Illinois's Urbana-Champaign Library in the late seventies. As reported by Mitsuko Williams and Elizabeth B. Davis of that library, "this program was highly effective in teaching biology students the use of the reference and bibliography collections in the Biology Library...."[5] The researchers found this technique to be thorough, convenient, and efficient—a great deal more so than traditional classroom, lecture-based instruction. The only extensive costs, aside from installation, phone lines, and computer time for a PLATO terminal, was the approximately $1,000 spent for programming.

Although this experiment included only seventy-two students, it does seem to have some value as an indication of the possibilities of such instruction. As compared to a control group, using traditional methods, the performance [getting "correct" answers] was very much better for the CAI group. Over 85 percent of the experimental group, as compared with 54 percent of the control group, "believed that the lesson would help them use the library." And this is only one example of PLATO's rather widespread use (estimated in the late seventies at reaching over 4,000 teaching stations). It is costly—approximately $15 per hour per student in 1976—but still less expensive than the estimated $20-$30 per hour for graduate students, and $30 and up for professional students, which traditional education costs. Another rather widely used form of CAI has the somewhat intriguing acronym of TICCIT (Time-Shared Interactive Computer-Controlled Information

Television). First developed by the MITRE Corporation (obviously, *M*assachussetts *I*nstitute of *T*echnology based), it was first used by community colleges but has already spread to some four-year colleges and particularly to Navy training programs in California and Florida.

Of course, the coming of the comparatively inexpensive stand-alone minicomputer has brought some thinking about its use as part of CAI. Millions of dollars worth of mini-computers are already in schools, from the elementary to the university levels, and their use will certainly affect the whole CAI picture. So far, what has been said about educational technology has not particularized its relation to the library. But it should be obvious that the audiovisual possibilities for schooling on all levels are fantastic. E. M. Buter has offered a rather inclusive list of educational media (all already available or soon to be as of 1977), which indicates the breadth and scope of already available educational technology.[6] He ranges through "instructional media and instructional television, including off-air audio or video recordings; telephone recording ...; audio-cassette; dial-up audio learning, and CAI (computer-assisted-instruction) at home."[7] This heralds a whole new role for the library.

## ACADEME, THE LIBRARY, AND *ACCIDIE**

Depending usually on your age, you will be conservative, liberal, or radical in your views on the salubriousness of the not-too-distant future of the academic library in America. The radical—usually the neophyte—will assume radical changes in both academe as a whole and in the library's part in it. Indeed, the more or less bland assumption, if you accept the computer-fanatic's enthusiastic prediction, is that both the university as such and the library as we know it will be greatly— perhaps almost unrecognizably—changed within a quarter-century or less. The conservative view—supported by Reaganomic cuts in federal support/aid for higher education, by inroads of anti-intellectualism, by the demographics of post-World War II U.S. population, and by some idea of just how little the fundamentals of American higher education have changed in the past thirty years—prefers to maintain the status quo in all aspects under consideration.

As is so often the case, the outcome of the years between now and the early twenty-first century will likely be somewhere in between the radical and conservative position. Or, if you accept the prophecies of the establishment futurists, then perhaps a little left of center is your best window on the future.

Harold Orlans, as part of the prestigious Commission on the Year 2000 of the American Academy of Arts and Sciences, some years ago predicted that "within the world of higher education, a progressive differentiation of institutional function can be expected to proceed together with closer local, regional, and national linkages of institutions and scholars.... Institutional collaboration will, of course, be facilitated by developments in the communication, recording, recall, and transmission of information in oral and visual forms."[8] He is cautious enough to let his

---

*A version of the material in this section first appeared in *Catholic Library World* (March 1982): 344-46.

predictions stay right there—in a somewhat generalized, non-detailed vision which still serves to pique one's curiosity as to actualities.

Orlans cites what, both in the sixties and now, seems like a practical goal for the year 2000—"the storage in machine-recoverable form of the entire deposits in the Library of Congress, and the accessibility of any item, without queuing, to readers at electronically-linked metropolitan and university sub-stations."[9] With such facilitation of scholarship and research as more or less "given," perhaps the most constructive use of time here might well be to deal with problems not so comparatively easily solvable.

These are the problems of the parent institutions to which the academic library is merely ancillary or service-giving. The academic library in the next quarter-century will rise or fall, improve or decline, as does higher education in America. And higher education *is* in trouble.

## HIGHER EDUCATION PROBLEMS

The most pressing problem, of course, is money. The Reagan cuts in both direct and indirect aid to colleges and the universities have been described by John Brademas, for many years the leading Congressional figure in working for educational improvement via federal aid and now president of New York University, as "so massive that they amount to a rewriting of Federal policy toward higher education."[10]

Indirect aid to students is also changing in a revolutionary, not evolutionary, manner. Publicly supported institutions whose political sponsors will have to find tax money to replace formerly available federal funds are already experiencing severe budget cuts. And private colleges that depended to a great extent on federal grants of one sort or another are bound to suffer. But somehow—even through the Great Depression—American higher education managed to survive. What seems a more basic problem than financing for higher education is a new attitude within higher education itself. That attitude is, in simple terms, a sell-out to expediency. Fortunately, some corrective measures have been suggested.

Thousands of thousands of acres of good timber have been sacrificed for the paper to print thousands and thousands of copies of reports now moldering in uncounted libraries and files. But one of the most cogent and accurately prophetic ones was the result of a cooperative effort in eight Western states, which in 1967 produced *Implications for Education of Prospective Changes in Society*.[11] Two of the more perspicacious chapters deserve special notice.

Paul A. Miller summarizes the "imperatives" for the seventies (and equally applicable to the eighties) as "urbanization and its rewards, work and its preparation, creativity and its nurture, morality and its strengthening, tension and its democracy and its workability, resources and their conservation, leisure and its use-time budget, [and] humanity and its improvability."[12] That litany is as useful today as when it was written a half-generation ago. He envisions the library of the future as a communication center with tapes (video as well as audio), microfilm thermoplastic retrievable storage, programmed learning units, and information recall systems. These items, he wrote, "are available not only here but to individuals at home and in areas far away. Yes, we find that books are still there, too. Many of them are coordinated with sight and sound."[13] Not bad haruspication for 1967, considering that silicon chips, the use of satellite relays, and so on were not yet even on the drafting tables!

And Gordon I. Swanson, in "Education for the World of Work," is equally vatic in his comprehensive view of the high future importance of adult education, vocational schools, and community colleges in a world where the imperatives of population growth and related demographic change, along with rapid technological progress, are certain to affect post-secondary education. He sees educational planning as critical in this area, urging that it be the result of combined endeavor by labor, industry, business, government, and (of course) educational leaders.[14]

## LIBRARIES, TECHNOLOGY, AND GOVERNMENT

In light of the sanguine and solipsistic views of the place of the academic library in higher education of the future that are commonly heard nowadays, one could almost—but not quite—forget or overlook the single greatest truth about that higher education. The fact is that it does *not* move rapidly toward change, even if that change is clearly for the good of the people being educated. The Carnegie Commission's 1972 report, *The Fourth Revolution: Instructional Technology in Higher Education*, dealt with the role of communications and data-processing technology in instruction and was rather precise in its forecasting. For example, it stated that "by the year 2000 it now appears that a significant proportion of instruction in higher education on campus may be carried on through informational technology—perhaps in a range of 10 to 20 percent."[15] Indeed, for the academic library, an optimistic chart indicated that electronic technology (computers, cable television, videocassettes) would be "generally introduced" by 1980.[16] Look about you. Have these items been "generally introduced," or is there even a strong prospect of their being so, as early as 1984? As for the small college, the report predicted that "students in small colleges will have more access to a greater variety of courses and greater library resources."[17] No supporting evidence is available to indicate the achievement of this utopian goal, either.

Perhaps one explanation for this debacle lies in another recommendation of the Carnegie study. This stated, unequivocally, that "the federal government will need to provide not only the bulk of the research and development funds, as it has in the past, but also funds for distribution of effective instructional programs that use the new technology ... [funds] equal to one percent of total national expenditures on higher education."[18] Unfortunately, the latest indicated trends come nowhere close to this recommendation. When billions are being slashed in an "across-the-board" fashion from domestic programs of all types, one can rest assured that education will also suffer.

As recently as 1980, the Research and Policy Committee of the important (and generally considered middle-of-the-road) Committee for Economic Development called for "substantial increases in federal funds for support of basic research at the nation's universities...."[19] They stressed that the source of socially valuable innovations was high-quality basic research. They held that such increased funding deserved a very high priority in future federal budgets. The present prospects for such priority are not very bright, as we all know.

The final report of the seven-year series of studies of American higher education by the Carnegie Council on Policy Studies in Higher Education predicted that "a downward drift in quality, balance, integrity, dynamism, diversity, private initiative, research capability is not only possible—it is quite likely."[20] But they were optimistic to suggest that this "drift" was "a matter of choice and not just fate." The key to avoiding the threatened disaster, in their view, was to manage for excellence not ineptitude—to accent "dynamism." And this is exactly where academic libraries and librarians can serve to the greatest advantage. The report—which by no means *features* libraries at any point—calls for "advancing the use of the new electronic technology ... [and] making better use of libraries as learning centers" as two of the "major new areas for initiatives," calculated to encourage the dynamism urgently needed to avert the decline posited to come within the next two decades.[21]

There is a political side to all this, of course, as politics is based on votes and votes are made by people. Figures indicate there are—in 1981-82, not the year 2000—over 32 million seekers of education beyond high school. This could be a very strong political force for obtaining the financial backing needed on every governmental level to give America dynamic, viable higher education.

Robert M. Hutchins, writing over a generation ago, described modern academe as producing "that strangest of modern phenomena, an anti-intellectual university."[22] He explains, in very simple terms, the place of higher learning in America:

> The universities are dependent on the people. The people love money and think that education is a way of getting it. They think too that democracy means that every child should be permitted to acquire the educational insignia that will be helpful in making money. They do not believe in the cultivation of the intellect for its own sake. And the distressing part of this is that the state of the nation determines the state of education.[23]

Clearly a nation such as the U.S. today, which politically has recently voted to denigrate service to the people and to go a-whoring after the gods of Mammon and Mars, is in trouble. So, education is in trouble. Hutchins claims that the only way to improve the state of the nation is through education; and education, he says, can only be improved when the state of the nation is such as to be willing and able to do so.[24]

## *ACCIDIE* AND LIBRARIANS

Now what about *accidie*? That now-obsolete term, popular in medieval times, was defined as "applied primarily to the mental prostration of recluses, induced by fasting, and other physical causes; ... the proper term for the 4th cardinal sin, *sloth*, sluggishness."[25] The sin of *accidie* is, if not currently pandemic in the world of higher learning, at least epidemic. It hardly requires documentation. At a time when the American college and university are struggling for their very existence, the American academic library community seems to concentrate on more and more demonstration of just how "mentally prostrated" they are, just how many of them are ivory-tower "recluses"—like Swift's Academy of Projectors' professors who concentrated on visionary technology, while "the whole country lies miserably waste...."

The academic librarian today should be doing everything in his or her (not too inconsiderable) power to convince academic administrators, boards, legislators, and the general public of the basic importance of "the Word." The storage and communication and wide dissemination of the world's wisdom (and folly!) are surely too important tasks to leave to the "information technician" and the burgeoning information industry by default.

Even that generally considered proponent of the medium as more significant than the message, Marshall McLuhan, said, "automation makes liberal education mandatory.... We are suddenly threatened with a liberation that taxes our inner resources of self-employment and imaginative participation in society."[26] He sees this as "a fate that calls men to the role of artist in society."[27] The academic librarian of the future must, then, be a liberated artist who takes a role in education as a *kochleffel*, a "stirrer-up," who carries on an unremitting struggle for the freedom and wide use of all available knowledge. *Accidie* is the American academic librarian's *first* cardinal sin ... (and its placid acceptance could mean sin's usual wages).

One of the wisest things that has been said about modern Western culture was the remark by Sir R. W. Livingstone that "the Greeks could not broadcast the Aeschylan trilogy, but they could write it."[28] The manifest progress we have made in technique is no substitute for our lack of basic creativity which today's obeisance to the Great God Career in our institutions of higher learning would seem likely to make permanent. In the eternal struggle between thought and action, thinking is usually a very distant second—and certainly seems likely to maintain that ranking today, as well as in the technology-dominated future now impending.

## CAI AND/VERSUS THE BOOK

Just as the book-periodical-pamphlet has been a backup resource for the traditional textbook-lecture education, so the library, by making available a wide range of supplementary software materials, and even by the loan/rental of necessary hardware, can combine its traditional book-based store of knowledge with what is newly available for what has been described, perhaps a little floridly, as "a polyculture of many interlocking teaching-learning situations."[29] The library then becomes an active, rather than passive, participant in the teaching/learning interaction. At first glance, this seems to contradict the vista of a Big Brother at one end of the teaching situation and the more or less captive student at the other end. But remember the saving grace of two-way audio, which gives promise of breaking the grip of what has been described as "frontal lecturing."[30]

Looking at this situation from another angle, how does the use of CAI, VDTs, and other ultra-modern library-usable technology affect the role of the librarian? If we accept Jesse Shera's dictum that "the aim of librarianship is ... to bring to the point of maximum efficiency the social utility of man's graphic records,"[31] then surely the librarian is only performing properly if he uses the machine or machine appurtenance that provides the aimed-at "maximum efficiency."

In this regard, Shera goes on to make a vital point about the distinction between information and knowledge. He stresses that "information is the impact of knowledge and is always received through the senses, no matter what or how many devices may intervene between transmitter and receptor...." Knowledge, he points out, " is that which an individual, a group, or a culture 'knows,' and there can be no knowledge without a knower." So, he concludes, "knowledge is everything an organism has learned or assimilated—values as well as facts or information—organized according to whatever concepts, images, or relations it has been able to master."[32]

Using Shera's valuable dichotomy concerning values and facts, let us then reassure ourselves that librarianship is not "selling out" in any way, and certainly not humanistically speaking, when it takes advantage of the nearly instantaneous communications and far-reaching scope that the computer and telecommunications combine to make possible. The librarian who uses the computer in any way is not—as should be obvious—thereby helping to "kill the book" or the many other possible uses of print. He is merely using the technological developments of his time—and, one would hope, using them with the best interests of his clients always under consideration.

The danger of over-concentration on means rather than ends is there. Of what avail is it, for example, to bring in relatively expensive CAI or other means of modern educational technology but to find the funds for doing so in starvation of book or periodical funds? The cost per hour of using a computer database once is still—and for years to come will be—almost astronomically more than will be the cost of buying a book, shelving it, and making multiple use of it through many years. And not only the simple cost/benefit figures need to be considered. When the MIT-minded instructional materials maestro begins—as some have already, in print and on the library school lecture circuit—to extol the virtues of the machine as if it will solve all the problems incident to our developing an information society, then beware the simplistic. Using Occam's Razor, that medieval device for resolving logical problems by seeking the simplest answer, is not always the best way to go.

What I am getting at is that all too often, when one is involved in a particular enterprise, one lacks the objectivity that can provide an accurate picture of the reality of that enterprise. Librarians, for example, are all too likely to look through the wrong end of the binoculars, so to speak, and to see their work as positively miniscule as compared to what it really is.

Perhaps a few facts and figures might provide a better perspective on this. Daniel Bell, in discussing what he calls "the post-industrial society," states that "the axial principle of the post-industrial society ... is the centrality of theoretical knowledge and its new role ... as the director of social change."[33] He sees what he calls "the crucial variables of the post-industrial society ... [as] information and knowledge."[34] The evidence for this is quite clear. During the half-century or so from the beginning of the Civil War to the end of the Philippine War (1860-1906), most American workers were on the farm. The next half-century involved the majority of American workers in industry. Since 1975, over one-half of all American workers were tied into the information society; indeed, by 1967, such labor earned over one-half of all work-derived income.

As databases, information networks, and the almost instantaneous international communications via multi-satellites bring about an even more accelerated growth of information and its retrieval than exists today, we may find an utterly new development in public policy. Contrary to what one might think, the new growth industry may be education. It is going to be very hard to keep the enterprising American from conquering the new "far frontier," that is to say, the frontier of knowledge—widespread, readily available, and, in the long run, relatively inexpensive.

# NOTES

[1] Carnegie Commission on Higher Education, *The Fourth Revolution: Instructional Technology in Higher Education* (New York: McGraw-Hill, 1972), p. 1.

[2] "Education and the Telefuture," *Change* (Nov.-Dec. 1979): 12-13.

[3] Anthony G. Oettinger, *Run, Computer, Run: The Mythology of Educational Innovation* (Cambridge, MA: Harvard University Press, 1969), p. 5.

[4] Ibid.

[5] "Evaluation of PLATO Library Instructional Lessons," *Journal of Academic Librarianship* (March 1979): 14-19.

[6] "The Great Shift—A Perspective on the Change of System-Resistance to the Spread of Educational Technology," in Philip Hills and John Gilberts, eds., *Aspects of Educational Technology*, v. XI (New York: Nichols Publishing, 1977), pp. 215-16.

[7] Ibid., p. 211.

[8] Harold Orlans, "Educational and Scientific Institutions," in Daniel Bell, ed., *Toward the Year 2000: Work in Progress* (Boston: Beacon Press, 1969), pp. 192-93.

[9] Ibid., p. 197.

[10] "For Universities, It's Harder Times in a More Complex World," *New York Times* (June 7, 1981): 21 EY.

[11] Edgar L. Morphet and Charles O. Ryan, *Implications for Education of Prospective Changes in Society* (Denver: Designing Education for the Future Project, 1967).

[12] Paul A. Miller, "Major Implications for Education of Projective Changes in Society," in Morphet and Ryan, p. 5.

[13] Ibid., p. 17.

[14] Gordon I. Swanson, "Education for the World of Work," in Morphet and Ryan, pp. 98-114.

[15] Carnegie Commission, *Fourth Revolution*, p. 1.

[16] Ibid., p. 1.

[17] Ibid., p. 4.

[18] Ibid., pp. 6-7.

[19] "More U.S. Funds Urged for Basic Research," *Higher Education and National Affairs* (Jan. 11, 1980), p. 3.

[20] Carnegie Council on Policy Studies in Higher Education, *Three Thousand Futures: The Next Twenty Years for Higher Education* (San Francisco: Jossey-Bass, 1980), p. 117.

[21] Ibid., pp. 103-4.

[22] Robert Maynard Hutchins, *The Higher Learning in America* (New Haven: Yale University Press, 1936), p. 27.

[23] Ibid., p. 31-32.

[24] Ibid., p. 32.

[25] *Oxford English Dictionary*. v. 1 (Oxford: Clarendon Press, 1933), p. 56.

[26] Marshall McLuhan, *Understanding Media: The Extensions of Man* (New York: New American Library, 1964), p. 310.

[27] Ibid.

[28] Hutchins, quoted on p. 25.

[29] Ibid., p. 62.

[30] Ibid., p. 63.

[31] In Edward B. Montgomery, ed., *The Foundations of Access to Knowledge: A Symposium* (Syracuse, NY: Division of Summer Services, Syracuse University, 1960), p. 9.

[32] Ibid.

[33] Daniel Bell, "The Social Framework of the Information Society," in Dertouzos and Moses, eds., *The Computer Age* (Cambridge, MA: MIT Press, 1979), p. 164.

[34] Ibid., p. 167.

# 4 LIBRARIES, INFORMATION CENTERS, and SCIENTIFIC PROGRESS

The distinction between libraries and information centers is, at best, a semantic one. The information center as generally described and operated is clearly *not* a conventional library. Basically, it is devoted to scientific information. It is also neither a so-called "specialized information service" nor a publisher. As far back as 1962, G. S. Simpson, Jr., rather definitively differentiated a scientific information center from various similar organizations, as follows: "Organizations of the following types are not 'scientific information centers'...:

1) specialized information services that have as their main objective the production of items such as selected abstracts, literature search, and bibliographical and accession lists.

2) specialized information services that mainly acquire, store, retrieve, and disseminate, upon request or on their own initiative, copies of their specialized collection of data or information [now called databases].

3) conventional scientific library.

4) publishing houses, trade associations, or professional societies that primarily print (not originate) and disseminate either scientific books or periodicals."[1]

## LIBRARY OR INFORMATION CENTER?

Granted these caveats, what, then, is an information center? Again, Simpson has as good a description as any available: "a scientific information center exists for the primary purpose of preparing authoritative, timely and specialized reports of the evaluative, analytical, monographic or state-of-the-art type. It is an organization staffed in part with scientists and engineers and, to provide a basis for its primary function, it conducts a selective data and information acquisition and processing program."[2]

A librarian would probably be part of such an "organization staff," in order to take care of the "acquisition and processing program" described above. But the raison d'etre of the scientific information center is very obviously quite different from the customary basis for a library. The preparation of reports is the frosting on the cake in a very few libraries, but it is certainly not a typical, normal operation. According to Simpson, there were "over 200 organizations (as of July, 1961) in the United States considered to be scientific information centers."[3] In the 1980s, the genre is ubiquitous, replacing, in a great many cases, the old "special library."

But the purposes remain the same, and thereby hangs a tale. Just how and why did the information center at least begin to replace the special library? It seems to be a combination of several factors, and it is highly debatable as to which came first. Simpson summarizes this, rather neatly, as follows: "Scientific information centers are, and will remain necessary to our continued scientific progress. Conventional libraries were sufficient up to the 20th Century. Specialized libraries then developed to supply services not available from conventional libraries. Scientific-information centers are but an extension of that trend."[4]

Now, why should such "think-tanks" (which, by the way, is *not* exactly synonymous with our description of all information centers) have developed around scientific/technological needs? The major reason for this is the peculiar or special nature of scientific research methods.[5] Where the humanist is usually a creative, people-centered researcher, where the social-scientist is normally equally people-centered, if less creative, the research scientist, per se, hardly ever follows "the common stereotype of the dispassionate, objective scientist at work in his secluded laboratory" being often as much or more occupied with fellow-scientists as with cold, hard facts.[6]

The main point, as Garvey notes, is that scientific communication is *interactive*.[7] It is not simply a matter of reading a book or performing an experiment; it is much more significantly a matter of disseminating tentative conclusions to a knowledgeable audience and using the resultant feedback to develop more refined, more accurate conclusions. This, of course, is really the explanation for the exponential growth of scientific journals in this century. The journal is the screen that shows the picture of what the scientist presents. But the trained reaction of the viewer matters almost as much as the original research. Under these circumstances, Garvey reminds us, the achievement of a particular scientist "will not be recognized as a scientific contribution until information about the phenomenon he has observed or discovered is integrated within the flow of scientific information, into the body of established knowledge in the field."[8] So it is scientific information *interchange* which really matters.

Therein lies the value of the scientific information center. As contrasted to the standard library, it is much more au courant with the latest data and information important to not-yet-published research. It is, as one might expect, quite a blow to any scientist to find that supposedly fresh conclusions are really old hat or negated by data with which he or she is not familiar. Variance in opinion is one thing; priority of discovery or ignorance of facts are quite other things.

For the purposes of this book, it is perhaps most important, to examine the consequences of the way scientists work as it relates to the freedom of scientific inquiry. A large proportion of American information centers are federally supported.[9] Those under the auspices of the Department of Defense are not normally accessible to non-defense-related scientific researchers.[10] So, in a manner of speaking, information is denied via secrecy classification. Just as one example, the Metals and Ceramics Information Center which is part of the Battelle Memorial Institute, has as its primary function "to provide information on the characteristics and utilization of the advanced metals and ceramics to those working for or on behalf of the Department of Defense and to industry."[11] And by "industry" is meant those firms with defense contracts, exclusively. There is nothing at all wrong with this, from the standpoint of safeguarding information vital to national defense and security but it is germane to the general question of how far freedom of scientific inquiry should go.

## THE WEINBERG REPORT
## AND LIBRARIANSHIP

Early in 1963, the President's Science Advisory Committee, using a distinguished panel of scientists and technologists headed by Dr. Alvin M. Weinberg, then director of the Oak Ridge National Laboratory, issued a truly prescient report entitled, *Science, Government, and Information: The Responsibilities of the Technical Community and the Government in the Transfer of Information.* One of the major points made by this report was that "mechanization can become important but not all-important."[12] Also, they saw "the specialized information center as a major key to the rationalization of our information system" and emphasized that "the specialized information center should be primarily a technical institute rather than a technical library..., led by professional working scientists and engineers...."[13]

It is notable that, in Samuel Goldwyn's famous phrase, librarians as such were mostly "included out." Indeed, in a brief discussion of the information process as part of the research process, the plain statement is made that "communication cannot be viewed merely as librarians' work...."[14] Over and over, the report stresses that the scientist of the future must be his own librarian; he must "compact, review, and interpret the literature...."[15] So, not the library, but "the specialized information center [is seen] as one key to ultimate resolution of the scientific information crises."[16]

Stating unequivocally that "our primary concern is to maintain the strength of our science and technology" (surely a worthwhile aim!), the panel says, "we must search for the means by which we can improve the efficiency of our communication system without sacrificing the values inherent in our traditional methods and organizations."[17] There seems to be some contradiction here between stated purposes and suggested means—unless, by some odd quirk, the library as we know it (hardly changed from 1963 in basic organization and techniques) is considered somehow not to be a part of our traditional methods and organizations.

One significant warning in this report is that "administrators and documentalists will have to improve their grasp of modern information-handling technology so that they do not look upon elaborate and expensive computers as magical panaceas for their information-handling woes."[18] All too often, the all-powerful, all-knowing machine has been considered in just this fashion, as some colossal failures of attempts at premature or ultra-automation of library processes have evidenced. Librarians may well be grouped with "administrators and documentalists" in this regard.

On the subject of teaching how to handle information, the report makes two rather helpful (for librarians and libraries) points: "accreditation teams should ... inquire not only into the adequacy of the library, as in the past, but also into the ways in which its use is promoted and facilitated ... Government agencies supporting research at a university should recognize support of the library as a legitimate expense." And then the report recommends that "more able scientists and engineers [should go] ... into technical librarianship."[19]

One of the greatest contributions of this report was its lucid, cogent description of the specialized data and information center. Note well that such a center is for *both* data and information. The SIC, as the report sees it, began as a data compiler "as opposed to ideas or knowhow"; its evolution was into not the disseminator and retriever of information, but into the source, the creator of new information, based on the data it had compiled. What those who work in such centers do is "to collect relevant data, review a field, and distill information in a manner that goes to the heart of a technical situation...."[20]

The report specifically enjoined that "new information centers be established at public and private technical institutions"[21] and "not as adjuncts of general libraries, or of publishing ventures, or of central depositories."[22] The basic reason for the clear differentiation, in modality and use, of information centers from libraries, says the report, is that the information center places its "emphasis on retrieval of information as contrasted with retrieval of documents...."[23] The vision of the authors (writing in 1963!) in foreseeing the current—and increasing—emphasis on retrieving information, not documents, is notable. Throughout the Weinberg Report, the stress is laid on "two primary and simple points: first; that information is part of research and development; and second, that *all* those involved in research and development—individual scientists and engineers, private institutions, industry, and government agencies—must become information-minded and must devote more of their resources to information."[24] Now that we are in the 1980s, it is apparent how very right the authors were in their judgments.

Although some of their more esoteric recommendations and prophecies—such as the one that called for direct government subsidy of secondary media (to the tune of $30 million annually by 1970)[25] or that which called upon authors to "refrain from unnecessary publication"[26]—are not ever likely to become realities, their final suggestion deserves as much—or more—attention now as it received in 1963. They gave as a "must" that the federal information system should do what it had to do but should not "overwhelm" nongovernmental systems, and that the two systems develop into "an effectively interwoven instrument that is always responsive to the changing needs of our science and technology."[27]

I know of no single government publication dealing with information/ communication issues that says so much in so brief (51 pages) a space. It deserves to be read by everyone concerned with the future of information technology and dissemination in the United States, both for its widespread influence and for the lingering echoes of its magisterial pronouncements, still thundering through the halls of both governmental and nongovernmental information-linked institutions. We are all in debt to Weinberg and his twelve fellow panelists.

Joan M. Harvey, writing in 1976, stated that the Weinberg Report "gave considerable impetus to the establishment of specialized information centres (or information analysis centres) and centres have proliferated since its publication."[28] This was not an effect limited to the U.S., since scientific information centers have been developed in rather large numbers in Great Britain and, indeed, throughout Europe. There are about fifteen so-called "information dissemination centers" in Europe, which deserve some brief attention.

## THE INFORMATION DISSEMINATION CENTER

The information dissemination center (IDC) is a specialized variety of information center that provides "commercially available current awareness, selective dissemination, and retrospective literature search services by searching the computerized data bases generated in the production of abstracting and indexing services."[29] Considering such abstracting and indexing services as secondary information sources, the IDC becomes a tertiary source. Such centers, of course, result from the exponential increase in the amount of material that needs to be abstracted and indexed.

## SCIENCE AND THE INFORMATION EXPLOSION

It has been said that the task of science is to create organized confusion out of chaos. As always with satirical quips, there is some substance to this one. The scientist, faced with a world full of data and information,[30] must select—or have selected—that which is pertinent, useful, or demonstrably true concerning what he or she is trying to prove. We are told that journals have increased at a rate of ten every fifty years since 1750.[31] Meanwhile abstracting services have expanded by ten every thirty years since 1860, and computerized indexes, at the rate of ten every ten years since 1949. The situation almost is *beyond* chaos!

There is, however, some light in this cloud of statistics. In an unusual study of the limits to growth in worldwide documentation, G. K. Hartmann urges that scientific information be better coordinated and integrated to avoid further needless duplication of effort the world over. He sees libraries as a filtering and coordinating mechanism toward that end.[32] The semi-independent information centers only produce more and more about less and less; libraries practice an economy of means and ends that helps the researcher to separate wheat from chaff. Alan M. Rees has offered some clarification of the dichotomy between libraries and information centers by stating that "the essential differences between librarianship and the newer concepts of information handling relate more to the type and extent of information services offered to the user than to the techniques employed to describe, store, and retrieve documents."[33] In other words, the information center, with or without the computer, is quite a different organization than the traditional library, and it needs to be recognized as such.

Rees' outline of the major ways in which the information center differs from the library is somewhat the same as Simpson's (cited earlier); he sees these four main points:

1.  Degree of delegation on the part of the information requester.

2.  Exercise of judgment and evaluation as to the significance and merit of the identified documents relative to the information needs of the requester.

3.  Provision of information and not documents.

4.  Processing of search output into a variety of search products—state-of-the-art review, critically compacted data, digests, and so on.[34]

If librarianship does not want to "permit the abnegation of the traditional function of the librarian—to satisfy the information needs of clientele," it simply must train the librarians of the future to be "concerned with the total information problem and ... [to] be as expert in the evaluation, selection, disseminating and presentation of information as in the techniques of storage and retrieval."[35] That's quite a tall order, especially if combined with librarianship's other classic aim, to make all this information as widely and freely available as humanly possible.

## SCIENTIFIC PROGRESS AND SCIENTIFIC INQUIRY

The rate of progress in science is directly related to the amount of freedom from limits a society places on scientific inquiry. A circumscribed, severely limited scientific community cannot possibly produce the quality and quantity of scientific research possible to free (or as nearly as practicably so) science. Whether it is DNA research, studies of the relative importances of heredity and culture/environment to human intelligence, nuclear power/weapons research, laser optics, biomedical research using fetal material, or experimentation with various chemicals on humans, there is no doubt that limits are now being—and will be in the future—imposed on American scientific and technological inquiry. And, of course, if scientists have limits, so do their right-hand assistants, librarians.

Perhaps the most practical way of looking at the collision course between societal limitations and scientific inquiry is to review the most thorough study done to date on the topic.[36] This far-ranging inquiry into inquiry was, for the most part, concerned with such biomedical research as the study of aging processes and, in particular, the possibility of use of recombinant DNA. But the principles, the opposing factions, are applicable to all scientific research and inquiry. In his introduction, Robert S. Morison deals with what he calls the "new anxiety" on four levels: "a concern for the harm that may be done to individuals in the simple pursuit of knowledge," "the concern for the possible damaging effects of new technologies that may result from new knowledge," "long-term hazards," and, finally, "the possibly unsettling effects of new knowledge on man's concept of himself and his relations to society or the rest of the natural world."[3][7]

He stresses what he calls the "social consensual" quality, which is science's only real justification for its work and its most basic requirement. Unless the knowledge that comes out of scientific inquiry is a *public* knowledge, a *shared* knowledge, Morison admonishes us, it cannot be truly science. "Reproducibility ... of the evidence upon which any new conclusion depends..." is essential.[3][8] Morison's other major point is that intellectual freedom is not only involved in writing and reading, but also in scientific experimentation and study. He defends "those remote creators of new knowledge" who render "undesirable acts" possible as not being responsible for the "benign" or "undesirable" uses of such knowledge. Indeed, he states, "the responsibility for the adverse effects would seem to be more appropriately borne by those who actually use the knowledge to commit undesirable acts...."[3][9] He concludes, "Society is almost certainly in greater danger from what it *does not know* ... than from what it *does know*...."[4][0] He wants an increase in the scientific community's willingness to take on inquiry that may be "boring and lacking in prestige," requires long years of study, demands a "high degree of cooperation between many different individuals and disciplines," or whose results "may make difficulties for influential vested interests in industry, or even in government itself."[4][1] There is, after all, just so much an individual scientist or a scientific center can do in finite time; the selection of priorities may be even more significant than overt, direct limitations of specific scientific inquiries of one type or another.

Perhaps the most important of all the essays in the *Daedalus* symposium now under discussion is Loren R. Graham's "Concerns about Science and Attempts to Regulate Inquiry."[4][2] What Graham deals with is the social responsibility of the scientific inquirer. He cannot treat this, he says, as one concern, but as a group of concerns about science. His "taxonomy of concerns"[4][3] includes a most interesting collection: "Destructive Technology," "Slippery Slope Technology," "Economically Exploitive Technology," "Human Subjects Research," "Expensive Science," "Subversive Knowledge," "Inevitable Technology," "Accidents in Science," "Prejudicial Science," and "Ways of Knowing."[4][4] Under each of these headings, Graham gives specific and useful examples.

The first, "destructive technology," includes such matters as potential and actual environmental damage that is an inevitable concomitant to the industrialization of modern civilization. Also included are some which are not purely environmental, "such as regulation of the distribution and use of pharmaceutical drugs, food additives and chemicals..., explosives, and radioactive materials." Control and regulation of "destructive" items (including military technology) are not seen by Graham as a matter "of principle about limitations on freedom of inquiry."[4][5]

Here I differ. When, as happened recently, a federal agency can force certain papers to be denied to an international scientific laser optics meeting because of so-called "potential" danger to national security, then the believer in intellectual freedom must be at least critical. The brand new idea in this is that there can now be, it seems, such a thing as post-issuance reclassification of material to a "secret" category, which denies information to foe and friend alike. This obviously offers a very dangerous precedent—but then, in 1978, Dr. Graham might not have foreseen a heated-up cold war as likely!

The next category of "concerns" dealt with in this essay is what Graham intriguingly called "slippery slope technology." Here the ethical scientist worries whether "in the biomedical field ... by blurring or erasing ethical boundaries that were earlier considered absolute, we will go out onto a 'slippery slope' of relativistic ethics on which we may lose our balance and tumble to the bottom." This is certainly not an idle fear. The ways in which Nazi "scientists" dealt with various almost incredible varieties of medical and eugenic experimentation should be lesson enough as to the, regrettably, almost infinite depths to which so-called "science" may fall. For this and other good and sufficient reasons, Graham does not see regulation in matters of this type as encroachments on scientific inquiry, but rather as necessary efforts to regulate possible abuses "while still taking advantage of the benefits of the new technologies, benefits which in many instances are significant and deeply humane."[46]

In his third category, "economically exploitive technology," Graham again sees little involvement in the scientific inquiry limitation area. He sees such matters as the possible development of an American supersonic plane or deciding on "the relative importance of research in curative medical technologies as opposed to improvement of environmental conditions" as being basically questions of social priority. He asserts, "if ... controls are carefully executed within a democratic framework, they are visualizable without danger to political or academic freedom." But now come a group of *scientific*, rather than *technological*, concerns. These, he says, involve "questions of principle that are essential to our concept of free inquiry."[47]

First among these is research on human subjects. Again, the actual occurrences in Nazi Germany, he stresses, warn us "against immoral and crude experiments." But even experiments that do not involve injecting deathly poisons or organisms into human guinea pigs can go further than some might see as morally justifiable. The coming of life-support technologies has clearly extended the moral responsibilities of the scientist/doctor far beyond what is commonly considered acceptable. Actually, what has happened is, in most cases, rather strict *self-*regulation. This, in a way, has broadened the limits of scientific inquiry because without such self-regulation, "the alternative is increasing regulation by bodies outside the scientific community."[48]

One of the most effective ways to limit scientific research is what Graham calls "expensive science." Who governs the allocation of the scientific dollar, whether from governmental or private sources, must, of course, govern what research gets done. With just so many dollars available, Graham says that "a worry about distortions in the allocation of resources is a serious one."[49] Is it a fundamental limitation on individual freedom not to be given a small grant because the money is being spent on a multimillion dollar, defense-related research facility? It is very hard to adjudicate such a problem.

In the category of "subversive knowledge," Graham includes these "classic" forms of what he calls "new fundamental knowledge": what conflicted with the accepted theories of current authority and what "appeared to demote the place of man in nature." Examples of each of these are the Church's censure of Galileo and Copernicus. Opposition to Darwinism in the nineteenth century combined both elements. According to Graham, currently rated "subversive" by some are studies in such fields as primatology, ethology, and extra-terrestial intelligence.[50]

What Graham calls "inevitable technology" is the doctrine that limits should be placed on basic scientific inquiry that "inevitably" will result in technological disaster—with fundamental physics research as the most commonly adduced example. Graham sees this as a "clearly false" argument, since many available technologies never have been and never will be put to actual use. It is impossible, he stresses, to foresee all the consequences of fundamental inquiry.[51]

The single striking example given by Graham on the concern about "accidents in science" is the argument over the presumed dangers of researching recombinant DNA. The City Council of Cambridge, Massachusetts, was so alarmed that it tried to prevent Harvard University by law from continuing such study.[52] Their immediate fear was "the possibility during fundamental research of the accidental production of a pathogenic organism immune to normal antibiotics."[53] There is, of course, justification for a local city council being concerned about possible threats to the public safety. Attempts to force "controls" over the distribution, storage, transportation, and use of chemically hazardous materials are becoming ever more common. But Graham sees the recombinant DNA controversy as most significant because of "(1) the authentic issues it raises about who should bear the responsibility of devising and enforcing regulations on research, and (2) the possibility that underneath the specific concerns expressed by the public on recombinant DNA look much more significant and deeper irrational fears of science."[54]

The next type of concern discussed by Graham, "prejudicial science," would really take volumes to examine thoroughly. Just to mention some of the subphases of science often included in this category—eugenics, intelligence-testing and its relation to race, and sociobiology—should indicate its high level of debate potential. Graham admits that this is one category where it is easy to see widespread "denial of freedom of inquiry in an area of fundamental research." He suggests that this will not happen if certain conditions obtain: "(1) no regulations are made prohibiting the research or the publication of the results; (2) fund administrators do not prohibit giving money for such research within general category fellowships and grants even if they refuse to support special projects of this type; (3) administrators do not interfere with the conduct of such research in institutions."[55]

Considering American librarianship's current strong united front on fighting censorship of all sorts, one might think that problems relating to "prejudicial science" would not be likely to arise. But, unfortunately, as recently as 1978, the leadership of the American Library Association, under pressure from its black caucus, attempted to pull back from circulation a movie—*The Speaker*—that the ALA Intellectual Freedom Committee had sponsored, which discussed, very peripherally, genetic theories of differential human intelligence. Fortunately, the ultimate governing body of the ALA, its Council, voted overwhelmingly to keep librarians from censoring themselves.[56]

With the coming of compunications, the perils of "prejudicial science" limitations seem to me to become more imminent and greater. If certain scientific data, whether "approved" or not, can even be considered as not appropriate for public, academic, or school libraries in any form or medium, such data could even more expeditiously be hidden or eliminated. The argument of a *New York Times* editorial is pertinent here: "the right of free speech rests on the premise that the airing of obnoxious speech is more beneficial to society than its suppression, that it is better for citizens to choose among contending ideas than for the state to do the choosing for them; that minority voices must be protected against the power of prejudice of the majority."[57] Graham has stated this dilemma very well: "a social environment so hostile to research of this type that no work could be done would pose, in fact, a true limit to inquiry of a sort that could be a dangerous precedent; on the other hand, a social environment in which certain political groups eagerly seized and successfully exploited arguments linking intelligence and race would present an extreme threat to society of another sort."[58]

The final Graham "concern" is with "ways of knowing." Under this heading he includes attacks on scientists and science itself by "creationists." He says, "one can only hope that the efforts to oppose religious, romantic, or mystical viewpoints to those of science will not receive significant official support."[59] To which I can only add, "Amen!"

Graham's conclusions on this whole matter of "concerns about science" are, as one might expect, rather middle-of-the-road. He admits that "scientists and engineers directly involved in the work have no monopoly of wisdom about the ethical, psychological, and societal impacts of their work." Yet, "at the same time we know that the assertion by lay groups of control over the determination of the inherent value of fundamental research could have disastrous results...."[60]

Personally, and as a longtime professional librarian, I lean to the side of open science as opposed to any form of closed society. As I said in an earlier work, "I, for one, cannot see how we who are members of a profession that is basically aimed at the widest possible distribution of information and knowledge can possibly support—even assist in raising!—barriers to such distribution or, more particularly, barriers to the discovery of knowledge."[61]

One other essay in the *Daedalus* symposium touches on fundamental issues on the topics discussed above. Dr. David Baltimore, a Massachusetts Institute of Technology microbiology professor and holder of a 1975 Nobel Prize in physiology, begins by stating, unequivocally, that "the traditional pact between society and its scientists in which the scientist is given the responsibility for determining the direction of his work is a necessary relationship if basic science is to be an effective endeavor." Society, in his view, "must determine the pace of basic scientific innovation," but "should not attempt to prescribe its directions."[62]

He is troubled in particular by the constant efforts of nonscientific groups to attempt to control research in basic biology, but he extrapolates this to all scientific effort. He gives two very pertinent, cogent answers to those who ask for limits on any research "because of the danger that new knowledge can present to the established or desired order of our society...." These are as follows: "First, the criteria determining what areas to restrain inevitably express certain sociopolitical attitudes that reflect a dominant ideology. Such criteria cannot be allowed to guide scientific choice. Second, attempts to restrain directions of scientific inquiry are more likely to be generally disruptive of science than to provide the desired specific restraints."[63]

He backs this up with two major considerations: "One is that science should not be the servant of ideology, because ideology assumes answers, but science asks questions. The other is that attempts to make science serve ideology will merely make science impotent without assuring that only desired questions are investigated."[64] He is against attempting to tell science where to go because non-science has no precision in its compass.

The limits of scientific inquiry should be as untrammeled as the limits on man's ever-inquiring mind. When an individual is told, "You cannot study this," whoever imposes the ban must first be omniscient. Until such an all-knowing censor arrives, essential scientific progress can only be delayed by the forces of ignorant censorship.

# THE CHALLENGE TO
# LIBRARY REFERENCE SERVICE*

There is absolutely no problem the computer cannot solve; there is no question the computer cannot answer. Just put the required information into the machine, and the machine will give it back, on request. But, unfortunately, none of this really solves problems or answers questions of reference service in the libraries of the United States in the next decade. These center much more on community access than on such substantive matters as the exact amount and placement of specific bits of information. Whether or not we achieve the so-called "paperless" library by 1990, as F. W. Lancaster has told us, we obviously cannot have any kind of library worthy of the name without patrons—and just who the library's patron, the consumer of information, is to be is much more in question than what information there will be and how it will be made available. Recent problems with state and local tax cuts have made clear that the library of today is not in the forefront of social priorities, and common sense would tell us that trend is not likely to change in favor of libraries.

If libraries are not going to be well (or even moderately) supported financially during the 1980s, among the first public services to go will undoubtedly be those proffered by reference librarians. And even before the individual ones vanish from public purview, their somewhat esoteric existence will be underfunded, if not completely undercut. The high-cost database access, the COM catalog, the videocassette—all far more expensive than print—may not be the cornucopias of goodies envisioned by the utopians. There is even some likelihood that librarians will be forced to go beyond the book to the person, to the local information source—the doctor, the lawyer, the banker, or even the local politician! Bill Miles has expatiated on this notion at length, saying, among other memorable things, that: "The library profession is not a profession that believes in getting its hands 'dirty.' We are probably as clean as a surgeon when we come to providing our services. This is unfortunate because information in a sense is not necessarily a clean subject. Information has to be dissected, sewn back together, and made operational. It has become bionic as well as neurotic in some degrees. It has to be part human and part mechanical to provide the kind of impact necessary to allow people to survive."[6][5]

It is the human, not the mechanical, side of reference service on which our hopes for a truly humanistic kind of heuristics should be based. No matter how fantastic the technological developments yet to come, they can only be fully exploited if the librarian of the future becomes more, not less, human. And by human, in this connection, I mean curious, involved—what a recent idiom called "with it." If I had my way, we wouldn't call reference service that at all. It should

*A version of this material first appeared in "The Challenge to Library Reference Service in the Decades Ahead," *The Reference Librarian* (Numbers 1/2, Fall/Winter 1981): 55-58. Reprinted with permission from The Haworth Press, Inc.

be "people service," because that's what reference librarians should be giving, now and in the unpredictable years to come.

The reference librarian of the 1980s must be more, much more, than a highly impersonal conduit to some vague thing called "information." Currently, only a very small proportion of libraries give more than a very small part of their reference service via providing access to online databases. But those who are using such services are finding the library's reference service to belie that appellation. Patrons are not being "referred"; they are not shown a book or handed a slip of paper with a call number on it or otherwise asked to "do it yourself." It is the librarian—new master of the modem and the bit, manipulator of the computer—who finds out for the patron what he wants to know.

Lancaster has predicted that by 1990, from 85-90 percent of all scientific and technical information will be available *only* as machine-readable data. There will be no way for the seeker after such information to get it without using the reference librarian as intermediary.

Fay Blake and Jan Irly have made a magnificent summation of just what the public library *could* do, vis-à-vis computer-related reference service: "it is important to recognize that public libraries can make vital use of the computer and the kinds of sophisticated services which the information industry makes available for business and industry. Freed from the limitations of the profit incentive, the public library could become the single most important community information resource, calling on existing data bases when appropriate, creating additional data bases of its own and appropriate to its own community, using technology for the maintenance and updating of current community information, and providing such extensive information without user charges as the right of all and not the privilege of those who can pay for it."[66]

This is a fine ideal—but the inevitable question of "who pays for the free lunch" will have to be answered. With budgets inevitably diminishing, just how will the public libraries of the nation pay the minimums, search fees, and line charges that are the concomitant of database use? And the setting-up costs of computer hardware are not going to diminish much, if at all, even though computers continue their miniaturization trend.

But I was going to address the human side of heuristics, which is perhaps best made clear by a little fable. Let us suppose that in the Land of Yrarbil the King is, of course, Nairarbil. King N has but one goal, one ideal, one faith: he wants to satisfy his royal consort, Queen Retupmoc. She is a very demanding, cold-blooded, perhaps even mechanical sort of individual. The more N does to please her, the more unhappy are his subjects, the great mass of the people, or, as they are often called, the Ytinummoc. Or, to put it algebraically, N loves R, R loves R, and Y used to love N—but soon will not.

Or, more directly, it is up to the reference librarian of the future to stay with the ideals of the last hundred years and think first of the community, not of the computer. Only then will the community be willing to support what really is best for its own preservation and benefit. The mutual gains from a librarian/community blending of funds and service should be both obvious and great. Finally, there are some current and continuing information technology developments that will undoubtedly affect reference service.

Within the past few years, a number of long technically available communication devices have become publicly available. In addition, information processing and disseminating devices have become linked in many previously unthought-of ways. The result of all this ferment is an angel's brew of developments likely to revolutionize completely such institutions as the corporation, the family, and the library. It is not exactly news to note that the computer, the satellite, television, and the telephone are rapidly becoming intermingled. Consider such developments as cable television, microprocessors, tele-text, videotape recording and playing, personal microcomputers with view screens, high-speed printers, and personal data banks—and you need not be an Edison or a Bell to foresee some likely revolutions in communication/information services within a decade or less.

The 1990s, it seems safe to say, will bring what Seth H. Baker, president of ABC Publishing, describes as a "home control center." This will include, he predicts, even a family ground station based on already available satellite access, combined with a cable that brings in more channels than even a "with-it" family could find time to view. In addition, this home control center, as he imagines, could "automatically perform mechanical functions such as controlling the environment, providing complete security against intrusions, and managing simple or complex home inventories." He goes on: "the family would also be able to do much of its banking, shopping, and bill-paying through the system. It would receive entertainment, mail, its equivalent of the newspaper, custom-tailored magazines for every member of the family, and information of all kinds—all by accessing varied services and data banks through the communications center."[6 7]

Baker predicts that there will be 20-30 million such centers in the 1990s. He doubts that the publishing industry will really be much different—except technologically—by then, but he worries that the media and their auxiliary enterprises need to be more seriously involved with computers and other new technology than they have been to date. And a late 1980 estimate in *Variety* claimed that by 1985, 25 percent of all U.S. households (25 million) will have either video-cassette tape recorders or videodisc players.[6 8]

The implications for the U.S. library of this prospect of exceedingly rapid mass-distribution of devices based on TV, the computer, and the satellite are many. The most significant for the reference librarian is that the scales seem well-balanced—the more information available, the more devices come to supply them. But it may well be the individual at home or in the office who will be doing his own reference work. And what will the reference library and reference librarian do then?

## NOTES

[1] G. S. Simpson, Jr., "Scientific Information Centers in the United States," *American Documentation* (Jan. 1962): 43.

[2] Ibid.

[3] Ibid. The form of appellation may vary, as Simpson's own usage indicates, but the concept remains the same.

[4] Ibid., p. 48.

[5] Discussion here is based mainly on William Garvey, *Communication: The Essence of Science* (New York: Pergamon Press, 1979), pp. 1-38.

[6] Garvey, p. 1.

[7] Ibid., p. 4.

[8] Ibid., p. 9.

[9] Joan M. Harvey, *Specialised Information Centres* (Hamden, CT: Linnet Books, 1976), p. 71.

[10] Some 56, of 200, hold classified information: Simpson, p. 47.

[11] Harvey, p. 72.

[12] President's Science Advisory Committee, *Science, Government, and Information: The Responsibilities of the Technical Community and Government in the Transfer of Information* (Washington, DC: GPO, 1963), p. 34.

[13] Ibid., p. 3.

[14] Ibid., p. 14.

[15] Ibid.

[16] Ibid., p. 15.

[17] Ibid., p. 19.

[18] Ibid., p. 21.

[19] Ibid., pp. 28-29.

[20] Ibid., p. 29.

[21] Ibid., p. 33.

[22] Ibid.

[23] Ibid., p. 34.

[24] Ibid., p. 50.

[25] Ibid., p. 37.

[26] Ibid., p. 25.

[27] Ibid., p. 51.

[28] Harvey, p. 13.

[29] Ibid., p. 85.

[30] Simpson [p. 43, ftn. 1 and 2] defines data as "number or letter patterns, recordings, and drawings which have relatively little meaning until interpreted," and information as "the recorded subjective interpretation of data, observations, or other information."

[31] Garvey, p. 7.

[32] G. K. Hartmann, *The Information Explosion and Its Consequences for Data Acquisition, Documentation, and Processing* (Washington, DC: National Academy of Sciences, World Data Center A, 1978).

[33] Alan M. Rees, "Librarians and Information Centers," *College and Research Libraries* (May 1964): 201.

[34] Ibid., p. 203.

[35] Ibid., p. 204.

[36] Gerard Holton and Robert Morison, eds., "Limits of Scientific Inquiry" (entire issue), *Daedalus* (Spring 1978): 1-236.

[37] Morison, "Introduction," in Holton and Morison, p. vii.

[38] Ibid., p. x.

[39] Ibid., p. xiv.

[40] Ibid., p. xvi.

[41] Ibid.

[42] In Holton and Morison, pp. 1-21.

[43] Ibid., p. 1.

[44] Ibid., pp. 2-3.

[45] Ibid., pp. 3-4.

[46] Ibid., pp. 4-5.

[47] Ibid., pp. 5-6.

[48] Ibid., pp. 6-7.

[49] Ibid., p. 8.

[50] Ibid., pp. 8-10.

[51] Ibid., pp. 10-11.

[52] Ibid., pp. 11-15.

[53] Ibid., p. 11.

[54] Ibid., p. 13.

[55] Ibid., pp. 13-15.

[56] A more detailed discussion of this topic appears in my *Defending Intellectual Freedom* (Westport, CT: Greenwood Press, 1980): pp. 48-62.

[57] "Two Celebrations of Free Speech," *New York Times* (June 11, 1978): 6.

[58] Graham, pp. 14-15.

[59] Ibid., pp. 15-16.

[60] Ibid., pp. 19-20.

[61] Oboler, *Defending*, p. 60.

[62] David Baltimore, "Limiting Science: A Biologist's Perspective," *Daedalus* (Spring 1978): 37.

[63] Ibid., p. 41.

[64] Ibid.

[65] Bill Miles, "From Prostitutes to Meter Maids—Unholy Sources of Urban Information," *RQ* (Fall 1978): 13-18.

[66] Fay Blake and Jan Irly, "The Selling of the Public Library," *Drexel Library Quarterly* (Jan.-Apr. 1976): 153.

[67] Seth H. Baker, "Foretoken," *Data Processor* (June/July 1980): 10-12.

[68] "TV to Homevideo, 1980-85," *Variety* (Sept. 24, 1980): 49.

# 5 MASS COMMUNICATIONS and INTELLECTUAL FREEDOM

*"Everything else is machinery; only freedom counts."*

— Herbert Agar, *The Perils of Democracy* (1965)

Sometimes a single brief sentence, a few words, can provide a truly illuminating insight into matters often obscured by the chatter of pundits. The coming of the machine to communication began, of course, with the fifteenth-century development of the printing press in Western civilization, but its seeming—or at least pending—triumph in our day has brought with it a tendency to overlook the key point of *all* communication: is it or is it not *free*? For all the centuries of Anglo-American law, there has been sharp and significant disputation as to the individual's right to know. The first real breakthrough on the British side of the Atlantic was when the Earl of Mansfield, Chief Justice of Great Britain, said: "to be free is to live under a government by law. The liberty of the press consists of printing without any previous license, subject to the consequences of the law."[1] This was elaborated, about the same time, by Sir Thomas Erskine, who, in arguing for permitting publication of Thomas Paine's *The Right of Man*, said (in part): "every man, not intending to mislead, but seeking to enlighten others with what his own reason and conscience, however erroneously, have dictated to him as truth, may address himself to the universal reason of a whole nation, either upon the subjects of government in general, or upon that of our own particular country."[2]

The end of licensing of the press was, of course, the beginning of true intellectual freedom in the United States. The First Amendment clearly and specifically enjoined the government from intervention in the freedom of the press. Although societal constraints necessarily bring limitations—for example, libel—we must always strive for the fulfillment of the traditional American/Jeffersonian idea, as expressed by Herbert Agar, "The press must be left free because it can help us keep an eye on government."[3] And the entire U.S. Constitution, constitutional experts agree, was devised for the purpose of safeguarding the individual from possible encroachments of unchecked governmental tyranny.

The very concept of "mass" communication has within it a concern, long before Marshall McLuhan so stated in print, with the medium of communication, the message to be communicated and, perhaps most importantly, who is the communicator and what audience(s) *receives* the message. Harold D. Lasswell has expressed this and more rather succinctly, as follows:

Who

Says What

In Which Channel

To Whom

With What Effect?[4]

As the communicator finds more and more outlets, an inevitably greater and greater opportunity exists for effect on thought and action. However, a whisper to a companion—if the whisperer is the President and the companion is one of the Joint Chiefs of Staff—might bring planetary destruction, whereas, a prize-fight seen the world over via satellite and films shown later might reach hundreds of millions—but would hardly matter in the eternal scheme of things.

It is of interest that when, in early 1982, the worldwide concern with the possible (even likely) use of nuclear weapons came to fever pitch, it was the printed word, rather than telecommunications, that stirred up a multinational following. In 1981 and 1982, nearly fifty books and countless magazine articles and news stories were published in English, each detailing the impact of a nuclear confrontation on both participating and innocent bystander nations.

But it is with telecommunications, in particular, that this chapter is mainly concerned. The most compendious and authoritative review of the likely effects of telecommunications saw it as "an area where the range of public policy and decision-making forms a web that links all parts of our economy and influences its effectiveness in serving our society."[5] It is, of course, true that the average citizen's direct relationship to telecommunications is mainly through the use of television, the telephone, and radio. But the vastness and ubiquity of telecommunications—most of which, one way or the other, affect importantly that same average citizen—is amazing.

Some of the major components of American telecommunications are as follows:

—Common carriers which give the public point-to-point communication service

1. Telephone

2. Telegraph

3. Teletype

4. Private lines

5. Data message switching

—Specialized common carriers: Communications common carriers authorized by the FCC to offer 'customized' private lines and data transmission services for business communications ...

—Value-added networks...: Common carrier entities that lease transmission facilities from other common carriers and add the value of
functions such as interfacing ...

—Miscellaneous common carriers: Communications common carriers
authorized by the FCC to provide video transmission services to such
customers as TV broadcast networks and CATV systems ...

—Domestic satellite carriers (DOMSATS): Communications common
carriers that operate communications systems in which microwave
signals are relayed between earth stations via the radio repeaters of
geo-stationary orbiting satellites....[6]

This is really a list of what are referred to as "structured components" of
the industry. In addition, services that come under the umbrella of telecommunications as currently defined include "private (intra-company or industry association)
networks, cable TV systems, and ... radio services—marine, aviation and air traffic
control, radio telephone, amateur, and numerous business and industrial systems."[7]
This is really the briefest possible outline of the telecommunications complex and
does not even refer to citizens band (CB) radio, probably the most ubiquitous and
fast-spreading of all types of communication in this country. Starting as a service
for the police, then extending to interstate truck service, and, most recently, to
be found in a great many private automobiles, CB is now accepted as of great
value in times of actual or potential emergency. A great number of CB licenses are
part of a so-called Radio Emergency Associated Citizens Team (REACT) described
as "volunteers ... linked to fire, police, civil defense, and Red Cross organizations
to provide emergency communications."[8]

The intellectual freedom consequences of the two-way mobile radio phenomenon are of some significance, as both free speech and privacy issues which have
been raised but not yet resolved. On free speech, the Little study asks (in part):
"Does the FCC have the right to dictate the duration, timing, or content of
messages transmitted over two-way radio? ... if it does have this right, can the FCC
possibly enforce it on a Citizens Band Service? Might the courts not view Citizens
Band Service differently than other two-way radio services, and give preference to
the rights of free speech specified in the First Amendment? Does the First Amendment really apply to CB?"[9]

Then they bring up the legality of police jamming of CB radio "to maintain
law and order in dealing with crime or civil disturbances." They question the
legality of confiscation (by FCC staff) of CB transceivers being used in clearly
illegal ways. Obviously, there is little, if any, privacy in two-way mobile radio.
Although mobile radio telephone systems are constructed so as to give call privacy,
CB is not. Does the owner of a CB radio ipso facto have the right to call privacy?
So far, the required technology has not appeared.

Very few, if any, clear and operative legal decisions have been made, even
below the federal level, on the many questions raised by the mass media's ubiquity.
The right of privacy (discussed, in a more general way, in chapter 4) has been ruled
by the courts "to exclude public officials and ultimately almost anyone in the
public eye on the grounds that their business is public if they are a part of the news
that is public, or on the basis that by becoming public figures, they forfeit the

right to privacy."[10] Clearly, this opens up a Pandora's box of difficulties. Does the accident of your being where a TV camera is aiming give the broadcaster the right to telecast your actions—or even your words? Or is there still some right that denies the newsperson's right to expose or feature you where and when you might not want to be exposed? The courts are really just beginning to get involved in firm, lasting decisions on such matters.

# INTERNATIONAL ASPECTS*

Macaulay described the 1695 abolition of prior-censorship in Great Britain by government licensing (a practice that had existed since a Star Chamber decree of 1538) as being of more importance for liberty and civilization than either Magna Carta or the Bill of Rights. This was not, of course, the end of censorship—but it made it a matter of judicial decision rather than of bureaucratic whim. Voltaire, Rousseau, and Benjamin Constant in France, Milton and John Stuart Mill in England, and countless others elsewhere throughout the history of man's attempt to practice freedom of thought and expression have vigorously advocated the principles on which, ideally, true intellectual freedom rests.

But countervailing opinions and actions have always calculated to inhibit such freedom. Classical Greece and Rome had their various forms of censorship—in Greece, against what was believed to endanger the state; in Rome, as Richard McKeon has put it, "protecting the native virtue of Rome from treason and immorality."[11] Christianity and Islam had their censorships, mainly of a doctrinal and academic type. The East, as well as the West, has a tradition of political censorship on a broad scale; in India, for example, for many centuries there was "imperial, censorial, and social suppression...."[12]

In fact, throughout recorded history, governments the world over have used censorship as a tool for self-preservation against sedition, to inhibit supposedly immoral or impious activities, and to control or deny the assertions of science where they were deemed irreligious or improper. In contradistinction to the First Amendment tradition in America, which puts the burden of proof on the censor, in much of the world outside of the United States, the censor (under a variety of names) is ascendant. Yet Western civilization has generally, since Gutenberg brought it printing, at least made a show of being discriminating and legal in its choice of what not to let its citizens say, read, and write. The continuing battle for freedom of the press is too well known and too lengthy to detail here. But by 1970, a knowledgeable British-trained librarian could say, with some wishful optimism but also with a great deal of favorable evidence, that "the trend in Western Europe appears headed toward complete intellectual freedom...."[13]

Unfortunately, much more prevalent than he expected, is political, governmentally approved censorship, not only in Western Europe, but throughout the world.

---

*A version of the following material first appeared in "International Aspects of Intellectual Freedom," *Drexel Library Quarterly* 18 (Winter 1982): 95-100.

Item—U.S.S.R.—The only way for Soviet creative artists (in all the arts) to get other than underground outlets is to satisfy the ubiquitous censors on every level—political, security (the KGB), and economic (professional artists' unions).[14]

Item—Brazil—Before each program is presented, Brazilian TV stations must show a certificate verifying that that particular program has been approved by the *Censura Federal.*[15]

Item—U.S.S.R.—According to Vladimir Borosov, expelled from the Soviet Union as a dissident in 1980, "our country has been information-starved for more than half a century. The government devotes considerable resources to prolonging the starvation."[16]

Item—Israel—Since Begin came in, Israeli television is "much more subdued ... censorship has become self-censorship ... there is far less criticism of the government's activities...."[17]

This is only a very limited sampling of a great many news stories, magazine articles, and books that indicate clearly that the Age of the Censor has far from passed. Although very few countries (as in Brazil) openly exhibit official "Censorship Departments," the work gets done, as in Soviet Russia, by means effective enough to manage to bar from public print, sight, or hearing whatever disturbs the powers that be.

The top researcher in this general field, Raymond D. Gastil of New York's Freedom House, has for several years rated all the independent nations of the world as to their relative state of freedom from governmental domination of the media. What he is referring to in these ratings, he says, is not "regulation such as that practiced by the FCC; government control means control over newspaper or broadcast *content.*" His most recent researches have indicated that "governments in three-fourths of the world have a significant or dominant voice in determining what does or does not appear in the media."[18] Sadly, only about a quarter of all the 161 nations surveyed keep *both* broadcast and print media generally free, while in about one-third, the press alone is relatively free. By his standards, "the press is partly free in twenty-four percent, not free in forty-three percent; broadcasting is partly free in twenty-two percent, not free in fifty-four percent of the nations."[19]

There is, as one might expect, almost a direct correlation between denial of intellectual freedom and denial of political freedom. The problem is that if the opposition is not given any opportunity to speak out without fear of punishment—ranging from imprisonment and torture to banishment and execution—the opportunity to vote hardly matters. As Gastil says, "freedom must include the right to be wrong, express foolish opinions, vote for poor candidates."[20] Viewing intellectual freedom's current status from another vantage point, there appears to be still another possible threat, which is international in scope. Since 1972, the United Nations Educational, Scientific, and Cultural Organization (Unesco) has been trying to develop a universal set of standards for the international news media, as opposed to what the United Nations set up as its model when it began in 1945—the free flow of information. A good many nations feel that the picture of what's going on in their respective bailiwicks is distorted by "capitalist" control of the major media—Reuters, the Associated Press, United Press International, the Agence France-Presse, and the American commercial networks. The "non-aligned" nations, in particular, resent the usual emphasis, as they see it, on disasters and catastrophes, on what is bound to look bad to viewers and readers in the developed, industrialized nations. So, in 1976, Unesco set up an International Commission on the Study of Communications Problems, chaired by Sean MacBride (winner of both Nobel and Lenin Peace Prizes).

This commission reported to Unesco in 1978, in a rather barouquely titled "Declaration on Fundamental Principles Concerning the Contribution of the Mass Media to Strengthening Peace and International Understanding, the Promotion of Human Rights and Countering Racism, Apartheid, and Incitement to War." This declaration, among many other things, forthrightly proclaimed that interference in the free flow of information violated a basic human right. It called for "diversity of sources and means of information, as well as the journalist's freedom to report, and fullest possible access to information."

This was not exactly what the Soviet bloc of nations sought as a formula. They called for a New World Information Order (commonly referred to as NWIO), since, they claimed, "Over-information and ... superabundance of the messages being transmitted are not tendencies to be fostered." In 1979, the Soviet Union's National Commission for Unesco held a conference at Tashkent, which described the free flow of information as a "grossly commercial concept serving the interests of transnational corporations...." During 1980-1981, a Unesco commission-sponsored conference in Belgrade indicated such substantial differences of opinion that no specific recommendations were adopted; another such conference was scheduled for 1983. Early in 1981, representatives of most of the non-Communist world's press organizations met at Talloires (France), under the auspices of the World Press Freedom Committee, and agreed to fight the proposals coming from Unesco, which they claimed would only result in tacit censorship.

This significant, wide-ranging dispute involves far more than just another blizzard of Unesco resolutions and reports. Such matters as whether "journalists would be obliged to promote government policies, subscribe to a code of information ethics, and receive identity cards that could be withdrawn if their work was given bad marks by Unesco officials"[21] were approved by the Third World majority at the Belgrade meeting. The restriction of data transfer across frontiers is another developing threat to the free flow of information.

Despite noble rhetoric in various recent constitutions and laws of some of the developing—and, of course, also in the developed—nations, the actualities of restriction of intellectual freedom internationally are much more than ominous; they represent to the 1980s what Turgot brought to the *ancien regime* in France, the censor at court. Whether accomplished by a United Nations agency or by individual government fiat, limiting intellectual freedom is a phenomenon of our times that verifies the perdurable value of Donald Thomas' perceptive comment, "The relevant question at any stage of human history is not 'Does censorship exist?' but rather, 'Under what sort of censorship do we now live?'"[22]

Just after the French Revolution, the Comte de Volney wrote a book on the possibilities of progress for the human race. He cited, according to J. B. Bury "two principal obstacles to improvement ... the difficulty of transmitting ideas from age to age, and that of communicating them rapidly from man to man."

De Volney claimed that the invention of printing removed these obstacles.[23] With our present capabilities for rapid, almost universe-wide communication—via satellites, teletext, computer-related machines of many types—it can only be man's seemingly inexhaustible capacity to find ways to promote censorship, to keep ideas from being transmitted, from being communicated, which lies in the way of true, universal intellectual freedom. Among the many significant tasks of the profession of librarianship, not the least is that of perpetual, unceasing awareness of and combat against censorship on every level, of every type, whenever and wherever it occurs.

## MASS COMMUNICATIONS
## IN DEVELOPING COUNTRIES

The relationship between progress in developing nations and the quantity and quality of mass communications equipment and its modernization is rarely considered in assessing the pace of world progress and prosperity. In a recent book, among other topics, Oswald H. and Gladys D. Ganley give at least a glimpse into the dimensions of this important topic.[24] They report that the U.S. Agency for International Development (in 1979) and the National Research Council (in 1978) called for rapid, early provision of improved communications to and from developing countries. But even such a development-minded global institution as the World Bank lent only a very small proportion of its resources for communications purposes.

The developing countries themselves have only devoted about .3 percent of their GDP (Gross Domestic Product) to improving their communications capabilities. This, the Ganleys tell us, "is less than half the average annual percentage of GDP devoted by developed countries," and they already have "major communications plants ... in place"! On a very basic level of international communications—the telephone—1976 figures show "developing countries, with 71 percent of the world's population and 19 percent of its income, had only 7 percent of its telephones."[25]

The Ganleys cite six reasons for the obvious great lack of support for communications and information in developing nations:

1. There is no existent infrastructure to receive it.

2. Communications and information systems are extremely capital intensive.

3. The economic benefits of communications and information have not been proven.

4. It may not benefit the poorest of the poor.

5. It benefits only urban areas and the elites.

6. Communications and information is political dynamite.[26]

It is the latter reason that infringes on the intellectual freedom area—or rather on its denial. If the government of a poor, developing country is either authoritarian or totalitarian (as most of them are), then, as the Ganleys say, "the rulers of the most needy nations may not be willing to rock the boat."[27] The simple fact of better informing the desperately poor of what they stand to gain under another form of government may touch off otherwise repressed rebellion or revolution. The 1979 Iranian revolution is perhaps the best-known modern case in point. The Ganleys hypothesize that the Shah "fell because he gave the Iranian people the communications and information resources which permitted them to know what they lacked and gave them the means to express their will."[28] That they moved from a political dictatorship into a theocratic one does not weaken this thesis.

# SATELLITES

As long ago as 1967, President Lyndon Johnson, in a message to Congress on communications policy, said, "communications satellites now permit man's greatest gifts—sight, expression, human thoughts and ideas—to travel unfettered to any portion of our globe. The opportunity is within our grasp. We must be prepared to act."[29] This all sounds very good—especially for intellectual freedom adherents. "Unfettered" is a good, rhetorical word—but what has happened to that impassioned presidential rhetoric in practice during this last decade and a half? In short—national security considerations.

As Rosemary Righter says, "Ownership and control has principally been used by both the U.S. and the Soviet Union for national and military purposes to date."[30] Although the United Nations long ago called for a "global satellite system to which all have equal access," such a system has never really come close to existence. Indeed, the reaction of a great many U.N. members—as stated in a Unesco declaration grandiloquently sub-titled "guiding principles for the use of satellite broadcasting for the free flow of information, the extension of education, and the spread of greater cultural exchanges"—was to call for a far different modus operandi and goal for "global" satellites. The declaration stated that "the objective of satellite broadcasting for the free flow of information is to argue the widest possible dissemination ... of news for all countries" and adjured that ' "cultural programs' ... should respect ... the right of all countries and people to preserve their cultures."[31] This was the first statement of a principle that has since come to be known as "controlled freedom."

In essence, "controlled freedom" calls for the right of any nation to license and regulate all members of the press. Since 1981, the U.S. Congress has stated as its policy that, if Unesco takes "any steps to implement a policy which has the effect of licensing journalists or their publications, to censor or otherwise restrict the free flow of information within or among countries, or to impose mandatory codes of journalistic practice or ethics, we in the United States shall withdraw our contribution to that organization."[32] How international freedom of communications can possibly exist under a licensure system has never been satisfactorily explained.

The democracies of the world have—as a group—responded strongly to the attempt by the mostly undemocratic Third World nations to force repression on the world's news and other communications. As mentioned briefly heretofore, at a 1981 meeting of some sixty leaders of independent news organizations from over twenty countries, held as a "Voices of Freedom" Conference in Talloires (France), a most important statement was issued. It deserves reprinting in full in this context, especially since it has not been widely reprinted elsewhere.

\* \* \*

## THE DECLARATION OF TALLOIRES[33]

We journalists from many parts of the world, reporters, editors, photographers, publishers and broadcasters, linked by our mutual dedication to a free press.

Meeting in Talloires, France, from May 15 to 17, 1981, to consider means of improving the free flow of information worldwide, and to demonstrate our resolve to resist any encroachment on this free flow.

Determined to uphold the objectives of the Universal Declaration of Human Rights, which in Article 19 states, "Everyone has the right to freedom of opinion and expression; this right includes freedom to hold opinions without interference and to seek, receive, and impart information and ideas through any media regardless of frontiers."

Mindful of the commitment of the constitution of the United Nations Educational, Scientific and Cultural Organization to "promote the free flow of ideas by word and image."

Conscious also that we share a common faith, as stated in the charter of the United Nations, "in the dignity and worth of the human person, in the equal rights of men and women, and of nations large and small."

Recalling moreover that the signatories of the final act of the Conference of Security and Cooperation in Europe concluded in 1975 in Helsinki, Finland, pledged themselves to foster "freer flow and wider dissemination of information of all kinds, to encourage cooperation in the field of information and the exchange of information with other countries, and to improve conditions under which journalists from one participating state exercise their profession in another participating state" and expressed their intention in particular to support "the improvement of the circulation of access to, and exchange of information."

Declare that:

1. We affirm our commitment to these principles and call upon all international bodies and nations to adhere faithfully to them.

2. We believe that the free flow of information and ideas is essential for mutual understanding and world peace. We consider restraints on the movement of news and information to be contrary to the

interests of international understanding, in violation of the Universal Declaration of Human Rights, the constitution of UNESCO, and the final act of the Conference on Security and Cooperation in Europe; and inconsistent with the charter of the United Nations.

3. We support the universal human right to be fully informed, which right requires the free circulation of news and opinion. We vigorously oppose any interference with this fundamental right.

4. We insist that free access, by the people and the press, to all sources of information, both official and unofficial, must be assured and reinforced. Denying freedom of the press denies all freedom of the individual.

5. We are aware that governments, in developed and developing countries alike, frequently constrain or otherwise discourage the reporting of information they consider detrimental or embarrassing, and that governments usually invoke the national interest to justify these constraints. We believe, however, that the people's interest, and therefore the interests of the nation, are better served by free and open reporting. From robust public debate grows better understanding of the issues facing a nation and its peoples; and out of understanding greater chances for solutions.

6. We believe in any society that public interest is best served by a variety of independent news media. It is often suggested that some countries cannot support a multiplicity of print journals, radio and television stations because there is said to be a lack of an economic base. Where a variety of independent media is not available for any reason, existing information channels should reflect different points of view.

7. We acknowledge the importance of advertising as a consumer service and in providing financial support for a strong and self-sustaining press. Without financial independence, the press cannot be independent. We adhere to the principle that editorial decisions must be free of advertising influence. We also recognize advertising as an important source of information and opinion.

8. We recognize that new technologies have greatly facilitated the international flow of information and that the news media in many countries have not sufficiently benefited from this progress. We support all efforts by international organizations and other public and private bodies to correct this imbalance and to make this technology available to promote the worldwide advancement of the press and broadcast media and the journalistic profession.

9. We believe that the debate on news and information in modern society that has taken place in UNESCO and other international bodies should now be put to constructive purposes. We reaffirm our views on several specific questions that have arisen in the course of this debate, being convinced that:

Censorship and other forms of arbitrary control of information and opinion should be eliminated; the people's right to news and information should not be abridged.

Access by journalists to diverse sources of news and opinion, official or unofficial, should be without restriction. Such access is inseparable from access of the people to information.

There can be no international code of journalistic ethics; the plurality of views makes this impossible. Codes of journalistic ethics, if adopted within a country, should be formulated by the press itself and should be voluntary in their application. They cannot be formulated, imposed or monitored by governments without becoming an instrument of official control of the press and therefore a denial of press freedom.

Members of the press should enjoy the full protection of national and international law. We seek no special protection or any special status and oppose any proposals that would control journalists in the name of protecting them.

There should be no restriction on any person's freedom to practice journalism. Journalists should be free to form organizations to protect their professional interests.

Licensing of journalists by national or international bodies should not be sanctioned, nor should special requirements be demanded of journalists in lieu of licensing them. Such measures submit journalists to controls and pressures inconsistent with a free press.

The press's professional responsibility is the pursuit of truth. To legislate or otherwise mandate responsibilities for the press is to destroy its independence. The ultimate guarantor of journalistic responsibility is to the free exchange of ideas.

All journalistic freedoms should apply equally to the print and broadcast media. Since the broadcast media are the primary purveyors of news and information in many countries, there is particular need for nations to keep their broadcast channels open to the free transmission of news and opinion.

10. We pledge cooperation in all-genuine efforts to expand the free flow of information worldwide. We believe the time has come within UNESCO and other intergovernmental bodies to abandon attempts to regulate news content and formulate rules for the press. Efforts should be directed instead to finding practical solutions to the problems before us, such as improving technological progress, increasing professional interchanges and equipment transfers, reducing communication tariffs, producing cheaper newsprint and eliminating other barriers to the development of news media capabilities.

Our interests as members of the press, whether from the developed or developing countries, are essentially the same: Ours is joint dedication to the freest, most accurate and impartial information that is within our professional capability to produce and distribute. We reject the view of press theoreticians and those national or international officials who claim that while people in some countries are ready for a free press, those in other countries are insufficiently developed to enjoy that freedom.

We are deeply concerned by a growing tendency in many countries and in international bodies to put government interests above those of the individual, particularly in regard to information. We believe that the state exists for the individual and has a duty to uphold individual rights. We believe that the ultimate definition of a free press lies not in the actions of governments or international bodies, but rather in the professionalism, vigor and courage of individual journalists.

Press freedom is a basic human right. We pledge ourselves to concerted action to uphold this right.

\* \* \*

This declaration has many implications for librarians, indeed for *all* individuals, groups, or professions sincerely interested in basic intellectual freedom. Especially notable is the constructive reference, in item 10, to the urgent need, in the field of communications/information for "improving technological progress, increasing professional interchanges and equipment transfers, reducing communications tariffs, producing cheaper newsprint, and eliminating other barriers to the development of news media capabilities."

There are already barriers enough to worldwide communications. Politically inspired "repressive tolerance," in the Herbert Marcuse/ultra-leftist style, which condones censorship in the name of repressive ideologies, cannot ever result in true intellectual freedom. As Congresswoman Fenwick said in discussing the Talloires Declaration, "The reason these concepts are so dangerous to the West is that they are based on a principle which is totally contrary to a fundamental tenet of Western society. The Soviet Union and many other nations in the world believe that the press exists as an instrument—a tool—of the State; that its duty is to promote the power and stability of the State. They do not believe, as we do, and our First Amendment proves it, that the press exists for the people."[34] So—may I add—do publicly-supported libraries.

## THE POLITICS OF INFORMATION

One of the most vital problems in dealing with freedom of information, worldwide, is that underlying the denial of free communications in many countries is an economic reason. As Joan Edelmaan Spero has put it, "recently ... the international information debate has taken a significant change in direction—from privacy and cultural sovereignty to economics."[35] What she is highlighting here is that the very fact of the ascendancy of information availability among factors leading to economic growth and technological advances has caused governments—both in developing and much more economically advanced nations—to shield their own information/telecommunications industries "by erecting barriers against transborder data flows—the flow of electronic information across national borders."[36]

Encoding messages vital to national security is one thing, but using exorbitant license and other fees, even denying the development of privately financed data networks to private companies obviously impedes, even stops, the free flow of information/communications of whatever type. Spero mentions, as a case in point, "newly adopted West German regulations [that] ... will prohibit users who are connected with the West German public network from transmitting all unprocessed data over private international leased lines to foreign private data banks for processing." In fact, "the national communications authority, Deutsche Bundespost, has also warned that beginning in 1985 users can expect to pay volume-sensitive charges on private international lines as well as on the West German public network."[37] Similar fees and charges are in force or planned in Canada, Brazil, and elsewhere.

No one claims that this is a vast, worldwide conspiracy aimed at preventing IBM or any other American data-processing firm from carrying on legitimate international business. But the results of such fiscal shenanigans will be—indeed, already are—paying what Spero refers to as "taxes on knowledge."[38] A few of the types of barriers rigged up recently by ultra-protectionist (for their own national telecommunications/computer industries) nations include: discriminatory pricing; local content laws ... requiring the processing of data within the country of origin; limits on the type of equipment available to international users; and ... restrictions on market entry. In addition, there are requirements for compatibility of technical standards, governmental monitoring of proprietory information, and even "outright denial of access to a national information system or entry into a market."[39]

Thus, under the guise of essential self-protection, many nations are, says Spero, challenging "the very essence of international production," with "direct threats to all forms of international commerce that depend on the unrestricted movement of data; to information businesses such as data processing and data base companies; to trade in services such as banking, insurance, and air travel; and to trade in goods that relies both on worldwide communications and on information-based services."[40]

What can be done about all this? Spero recommends "an effective U.S. tele-communication and information policy [that] will be responsive to a broad range of issues from privacy regulation to freedom of the press, to telecommunications policy, and to national security. In addition ... the United States must develop an active economic strategy that serves U.S. economic needs and preserves the principle of free flow of information."[41] As a vehicle for an international "institutional forum" that will develop a governing system based on rules agreed upon by the international community, Spero suggests GATT (the General Agreement on Tariffs and Trade).[42] Her point is that GATT is already functioning—and functioning very well—as "a flexible multilateral negotiating forum" and is dealing with the international flow of data is not beyond its capabilities or already recognized responsibilities.[43] She admits that the knotty problems incident to international compunications will not be solvable right now. In fact, she says, "it will be necessary to inventory barriers to information flows, to examine technological trends, and to explore the relationship of GATT's mandate with the responsibilities and authority of other international institutions."[44]

However it is done, an American policy on information is way past due. Or, as Spero concludes, "the United States must begin to assemble an information policy today if it wishes to avoid the high cost of unpreparedness tomorrow."[45] Remember that compunications are changing every day, and that their international effects are daily seen to be more far-reaching and complex. Stumbling along on a catch-as-can, quotidian "policy" is simply not good enough.

The international aspects of communication are complicated, of course, by political factors. When the bombing of West Beirut, Lebanon, was going on in the summer of 1982, not only were observers and reporters affected by the very fact of being on the spot during bombing, by the availability or non-availability of satellite or cable access, but also by political censorship. The Israelis, as might be suspected, wished to indicate their pinpoint accuracy in hitting only PLO-infested targets; the PLO wanted to show that Israeli bombing was simply an indiscriminate slaughter; the Lebanese were only concerned with getting the whole thing stopped. But censorship was evaded by the members of the free press, in many cases—so the truth (at least through each individual reporter's eyes) did reach the outside world.

# NOTES

[1] *22 State Trials*, p. 414—cited in Frederick Seaton Siebert, *Freedom of the Press in England 1476-1776: The Rise and Decline of Government Control* (Urbana: University of Illinois Press, 1965), p. 392.

[2] Siebert, p. 392.

[3] Herbert Agar, *The Perils of Democracy* (Chester Springs, PA: Dufour Editions, 1965), p. 21.

[4] Harold D. Lasswell, "The Structure and Function of Communication in Society," in Wilbur Schramm, ed., *Mass Communications* 2nd ed. (Urbana: University of Illinois Press, 1960), p. 117.

[5] U.S. Dept. of Commerce. National Technical and Information Service. *Telecommunications & Society, 1976-1991* (Springfield, VA: NTIS, 1980 [dated June 22, 1976]), p. 3.

[6] Ibid., p. 10.

[7] Ibid., p. 11.

[8] Ibid., p. 48.

[9] Ibid., p. 61.

[10] Frederick C. Whitney, *Mass Media and Mass Communications Society* (Dubuque, IA: Wm. C. Brown, 1975), p. 107.

[11] Richard McKeon, "Censorship," *New Encyclopaedia Britannica*, 15th ed., v. 3 (Chicago: Encyclopaedia Britannica, Inc., 1974), p. 1084.

[12] Ibid., p. 1087.

[13] Robert Collinson, "Trends Abroad: Western Europe," *Library Trends* (July 1970): 121.

[14] Serge Schmemann, "Commisars of Culture Don't Relax Very Often," *New York Times* (Nov. 8, 1981): 8E.

[15] John Wicklein, "The Long Shadow of Censorship," *Atlantic Monthly* (Aug. 1979): 13.

[16] Charles Allen, "Talking with a Soviet Free Trade Unionist," *New Leader* (Sept. 21, 1981): 8.

[17] Rivka Fried, "Begin Squeezes Israeli TV," *New Statesman* (Nov. 21, 1980): 12.

[18] Raymond D. Gastil, with Leonard R. Sussman, *Freedom in the World: Political Rights and Civil Liberties 1980* (New York: Freedom House, 1981), p. 73.

[19] Ibid.

[20] Ibid., p. 9. Most of what follows here is based on Gastil, pp. 54-98.

[21] Paul Lewis, "Gloves Come Off in Struggle with Unesco," *New York Times* (May 24, 1981): 3E.

[22] Donald Thomas, *A Long Time Burning: The History of Literary Censorship in England* (New York: Frederick A. Praeger, 1969), p. 318.

[23] Count de Volney, *Les Ruines des Empires* (Paris, 1879); quoted in J. B. Bury, *The Idea of Progress* (New York: Dover Publications, 1955), p. 200.

[24] Oswald H. and Gladys D. Ganley, *To Inform or Control? The New Communications Networks* (New York: McGraw-Hill, 1982), pp. 115-25.

[25] Ibid., pp. 116-17.

[26] Ibid., p. 119.

[27] Ibid., p. 123.

[28] Ibid., p. 124.

[29] Quoted in Joseph Newman, ed., *Wiring the World: The Explosion in Communications* (Washington, DC: U.S. News & World Report Books, 1971), p. 98.

[30] Rosemary Righter, *Whose News? Politics, the Press, and the Third World* (New York: Times Books, 1978), p. 219.

[31] Ibid., pp. 220-21.

[32] Robin Beard [Cong., TN] in *Congressional Record* (Sept. 17, 1981), p. H6348.

[33] Inserted into *Congressional Record* by Millicent Fenwick [Cong., NJ] (Sept. 17, 1981), p. H6354-55.

[34] *Congressional Record* (Sept. 17, 1981), p. H6356.

[35] Joan Edelman Spero, "Information: The Policy Void," *Foreign Policy* (Fall 1982): 139.

[36] Ibid., pp. 139-40.

[37] Ibid., p. 143.

[38] Ibid., p. 144.

[39] Ibid.

[40] Ibid., p. 145.

[41] Ibid., p. 152.

[42] Ibid.

[43] Ibid., p. 153.

[44] Ibid., p. 154.

[45] Ibid., p. 156.

# 6

## The CITIZEN and the COMPUTER

## Privacy, Freedom of Information, and Copyright

When computers first began to be widely used for the storage and retrieval of information—from the late fifties through the late sixties—most of what was involved was the transformation of already existing files into machine-readable files, for just about the same purposes for which the traditional ones had been gathered and kept. During the late sixties and the very early seventies, accumulation of data via the computer was very often for some specific purpose, rather than simply being based on one firm or bureau's records. Since then, data banks have been expanded into so wide an area that, as Brier and Robinson say, "files on individuals could contain comprehensive information and be available to a number of different agencies through an efficient computer system."[1] Here is where the greatest danger to individual or group privacy can occur.

Obviously, if all the details concerning the personal life of an individual available somewhere in some computer data bank can be pulled together almost instantaneously and made readily available, there could very well be a significant threat to individual privacy. The details forthcoming from data banks, just as obviously, would not necessarily be matters of fact, and there would be no way for a defamed individual to challenge likely-to-be-inaccurate statements. Simply gathering together what the FBI calls "raw data" is certainly not a refining or a fact-gathering process.

There are those who say that these possibilities for error make "the very existence of any form of data bank a likely or presumed violation on individual privacy."[2] What seems to be a more significant question is what rights to privacy an individual could (or should) have. Should there be a nationwide record kept of all arrests, for example, whether the individual is ever tried for the reputed crime or not? The FBI's National Crime Information Center, imperfect as it is because some local jurisdictions do not provide complete data of this type, is attempting to be a one-place file of this type—and there is absolutely no guarantee of either the accuracy or currency of what the FBI has available. And the information in the NCIF is *not* restricted only to law enforcement agencies. There are many proven incidents of such information's being made available on request to prospective employers—who may easily be misled into using FBI "raw data" as unassailable fact.[3]

Another example of privacy rights invasion, one of which resulted in an unsuccessful law suit, is the computerized record-file of the State of New York's Department of Motor Vehicles. This includes "license and registration files, insurance data, accident histories, scheduling of hearings for traffic violations, and medical information."[4] As if their compilation of such data were not enough, the NYDMV is, by law, permitted to sell lists of its vehicle registrations to the highest bidder, and it sells other related lists to insurance companies and bank/credit firms. There certainly would seem to be privacy violations involved in such dealings, although the courts so far have not agreed.

One of the most difficult areas of privacy violation is in the compilation of so-called "political files." I am not referring to the kind of computer lists maintained by legitimate political parties for the sake of fund-raising or opinion-gathering (or influencing). As far back as 1958-1959, a series of articles, mainly on this topic, appeared in *The American Scholar*,[5] written by a large group of distinguished litterateurs and journalists who minced no words in their strictures on what they felt was the "Invasion of Privacy." They saw this as bad then and likely to become worse in the future.

Richard H. Rovere, like the others, was writing before computerization highlighted the justice and omnipresence of his cause, but what he complained about as problems with technology then—wire-tapping and "bugging"—are equally worth considering in today's computer-age. He saw the basic causes of the erosion of privacy in two phenomena, "our advancing technology and in the growing size and complexity of society." With conditions no different in the 1980s, perhaps unexpectedly worse—it is no less true today than in Rovere's era that "technology has forced the surrender of a measure of privacy...."[6] We are, as Rovere reminds us, "perhaps the most gregarious and community-minded of people and have developed social and technological interdependence further than any other...."[7] Have we then created our own Frankenstein's monster?

August Hecksher thinks so also. He says that "the machine is used by man to further his own ends and it can protect privacy (if that is what we desire) as well as circumscribe it."[8] In other words, if we have problems with technology versus privacy, they are self-created. He calls for recognition that we have fallen "into a misty mid-region of social conformity."[9] Although this was written in the doldrums of the Eisenhower era, it seems even more appropriate for the Reagan period. Hecksher does not really blame government or the machine for what he calls "the reshaping of privacy." To cite George Meredith, "In tragic life, God wot, No villain need be! Passions spin the plot: We are betrayed by what is false within."[10]

A third symposiast, literary critic Granville Hicks, gives us a point of view not so much related to the problem of privacy invasion incident to the computer, but rather to the basic philosophical problem of privacy seeking. He sees no possibility these days of living in a community, great or small, and expecting *complete* privacy. As he says, "it is clear that a great society cannot function, any more than a small town can, without some invasion of the privacy of its component parts."[11] He sees *no* such things possible as the absolute right to be let alone. We are all too interconnected for that.

He points out that even such a critical issue as the willingness to admit publicly one's political beliefs (shades of the House Un-American Activities Committee and its threatened current successors!) hinges on a significant philosophical issue. Beyond the secret ballot—which he sees as sacrosanct—he stresses that "our political life would collapse if all men, or even if very many men, insisted on regarding their political views as a private matter." He goes on, "it is the individual who must make the decision; society must not coerce him." But he sees in all this that "the key word is not privacy but responsibility. A responsible individual will recognize that representative government is impossible without a free and constant interchange of opinions. He must be willing to state his views, even if they are unpopular, and stand by them."[12]

The final contribution to this important (pre-computer-age, mind you) symposium with which we have been dealing was journalist Gerald W. Johnson. What he—and I—are concerned with is the tendency toward a rule of the majority that ignores the rights of the minority. Putting this in context, does the invasion of the privacy of the vast majority of Americans—without any explicit granting of that right by them—mean that the minority cannot dispute this action? Most unfashionably—whether in 1958 or in 1982—Johnson decries what he calls "the egalitarian heresy that denies there is a rabble." Indeed, he criticizes what he calls "gutless intellectuals" who are, he says, "afraid not of a large enemy with a club but of loneliness. They cannot endure the solitude of the sentry on outpost duty.... They surrender the right of privacy because they lack the hardihood to take the position that what the rabble cannot understand it has no right to know."[13]

I realize this kind of attitude sounds "elitist," "aristocratic," "anti-egalitarian," even "un-American." But remember that Thomas Jefferson said not only that "the will of the majority is in all cases to prevail," but added, "that will, to be rightful, must be reasonable."[14] In these days, it is good to remember that, as Johnson puts it, "the will of the majority may not in all cases be rightful, and when it is not rightful it is contemptible."[15]

But perhaps this writer may seem to you to have gone far afield from the questions incident to privacy versus or despite the computer. I don't think so. Unless we think of the fundamental issue, the one behind the current machine-made dilemma, we may in truth be bowing down to Moloch, selling out to the machine. Or, as Richard Rovere said, "we are perhaps the most gregarious and community-minded of people and have developed social and technological interdependence further than any other, but it is still ... universally acknowledged that the man who tells another to 'mind your own business' has justice on his side and speaks the common law." Indeed, he goes on, "We are all in the same fix, and we all have to strike the same balance between our need for others and our need for ourselves alone."[16]

Strange, strange words ... aren't they? In this day and age, this kind of thinking is practically heresy. But what of that? Let's try to be heretics—devil's advocates—at least for the next few pages.

Let's suppose that somewhere, sometime in this benighted land of ours a librarian decides that all this networking, interconnecting, cooperation, joint effort, and onlining has too many dangers to justify doing the regional "dance of the computers." Let's further imagine that this dissident librarian is in charge of an academic library in a college with about 1,000 students and fifty to sixty faculty members, a library with some 100,000 volumes and 500 periodical subscriptions. The librarian refuses to give interlibrary loan service and asks for none. In other words, the total library resources available to the campus community are then those that the library happens to have on its premises or that it adds, on occasion.

I don't know about you, but *I* feel this *could* work—assuming that *only* undergraduate degrees are granted and that the faculty will *not* be required to "publish or perish." This is certainly not directly related to privacy considerations, but it is an example of the possibilities for non-conformity still possible in a world where the rate of technological advance may be just too rapid for someone who, as Henry David Thoreau said, hears "a different drummer."

To return directly to the topic of privacy, perhaps the foremost legal authority on the subject in America, Alan F. Westin, has suggested that what we have in the privacy problem is a bivalent, bipolar proposition. At different times, the same individual or the same society wants privacy and wants to invade the privacy of others; to be let alone to do as one wishes and (to use Westin's words) for society "to engage in surveillance to guard against anti-social conduct."[17] He sees America as "a democracy whose balance of privacy is continually threatened by egalitarian tendencies demanding greater disclosure and surveillance than a libertarian society should permit."[18]

In other words, we want to be free—but we have to admit that complete freedom—even as with complete privacy—is impossible. This dilemma puzzles some people. As Westin says, "many governmental and private authorities seem puzzled by the protest against current or proposed uses of new surveillance techniques."[19] Indeed, they ask, "aren't fears about subliminal suggestion or increased data collection simply nervous responses to the new and the unknown?" After all, the argument goes, if you're "clean" and really have nothing to worry about or hide in the public or semi-public exposure of yourself, why worry? Westin sees this as an academic question. As he says, "The right of individuals and organizations to decide when, to whom, and in what way they will 'go public' has been taken away from them. It is almost," he goes on, "as if we were witnessing an achievement through technology of a risk to modern man comparable to that primitive men felt when they had their photograph taken by visiting anthropologists: a part of them had been taken and might be used to harm them in the future."[20]

## COPYRIGHT AND COMPUTERS

Now—what about the copyright issue in years to come? As far back as 1908— in the Supreme Court case of *White-Smith Music Publishing Co. v. Apollo Co.*, the matter of adapting copyright laws built around print materials to the exigencies of the new twentieth-century reproducing systems has been a problem. The 1908 ruling was that phonograph records or piano rolls—any copies of musical compositions, not printed or written—were not safeguarded under then-existing copyright law. Actually, it was not until 1971 that a new federal law specifically named "sound recordings" as being under copyright. Motion pictures, radio and television broadcasts, photocopies—all have had their day in court and in the halls of Congress.

But what about the newest of the mechanical monsters, the computer? What about—to use the most commonly used term in this area—the validity and appropriateness of copyright in what are called "computer-readable works"? Without going into all the technical points concerned, here is essentially what came out of the most thoroughgoing study on the subject to date. Roy Saltman's *Computer Science & Technology*, prepared in 1977 for the National Bureau of Standards, gave a very effective and really encouraging aspect to what had seemed an almost inpenetrable thicket of laws and court rulings.[21] Saltman cites a 1967 report and agrees with its most startling prediction, that "the computer, in essence, assumes the role of a duplicating rather than a circulating library. One copy of a book fed into such a system can service all simultaneous demands for it...." Then, rather mildly in view of the implications of this dazzling statement, this prediction states, "of course this substitution for additional copies can vitally affect the publishers traditional market."[22] And, may I add, the librarian's, too!

Before the Copyright Law of 1976 was passed, Congress created a national commission (CONTU) to look into the increasingly pressing problem of new technologies and their copyright implications. This commission is continuing its study, with several reports already available. It seems very likely that what will happen with the 1976 Copyright Law will be—as in the case of such post-1906 technologies as the phonograph record, film, radio, television, and photocopying—that valiant attempts will be made to use court decisions, rather than legislation, to adapt statutes to each new situation. Certain questions will have to be asked, as explained by Saltman: "1. Is the new product copyrightable? 2. What rights are covered by the copyright in the new product? 3. Are new devices for using copyrighted works subject to the copyright?"[23] For motion pictures, sound recordings, microfilms, photocopying, and radio and television broadcasts, many issues arising from these questions have been resolved in judicial, legislative, bureaucratic, or the Office of Copyright's rulings. For computer programs, the road ahead is very murky.

Just as an illustrative side-issue, there is now a controversy in the semiconductor industry over what is called "reverse engineering" of silicon chip integrated circuits. A 1980 article in *Business Week* referred to "silicon spies" and cited as now in progress an attempt by the American Electronics Association, described as "a major industry trade group" to amend the federal copyright law to give "extended copyright protection to semiconductor circuits." There are big stakes in this matter; the article states that "the latest generation of integrated circuits can easily require an investment of upwards of $5 million and two or more years of design engineering."[24]

This may sound remote from library concerns, but consider this: in some not-too-distant future, will it become necessary for libraries to *copyright* their original catalog cards, book lists, or other products? This all seems very foreign to our long-established tradition of interlibrary cooperation; but we are beginning to be a part of a very large unit indeed—the information industry—and the time of laissez faire in library work may be passing more rapidly than some of us would like.

Saltman sees the computer program/copyright matter (as of 1977) as a problem for quite a group—authors, publishers, libraries, educators and students, research organizations, computer hardware and software producers, computerized information service system owners, and commercial indexing and data search services.[25] Saltman states that "data bases, whether in printed or machine-readable form, are copyrightable as compilations."[26] Therefore license/lease agreements will have to be made between database "publishers" (to use the least technical term) and system operators and users. No such license would permit the extraction of an almost complete copyrighted database, or even a major part of it, just as in the case of books.[27]

What about copyrighted books/documents/periodicals available only or also in machine-readable form? Saltman says that "the discussions and conclusions in this study relating to data bases are applicable generally to the computer storage and retrieval of the full text of documents."[28] In this regard, Saltman sees some justification in establishing what he calls "voluntary 'clearinghouse' systems" in order to give blanket licensing of the use of copyrighted works in various types of computer arrangements. Or else, he suggests, there might be a statutory system to require that such works in such systems be licensed.[29]

That is about as far as I would like to venture, at this time, into the quagmire of computer-product copyright. It is difficult enough to understand the Copyright Law of 1976, especially in its relation to libraries, education and research, and "fair use." It is to be hoped that, within the next few years, either a substantial amendment of the present copyright law or (as Saltman suggests may be better) a special statute for computer-related copyright would at least to some extent clarify the whole situation.

It might be of interest to know that a substantial portion of the computer industry (*not* including those who *use* their product, I might stress) is completely opposed to extending copyright to cover computer programs in any way whatsoever. Their main point is that programming is not analogous to book publishing. Philip Dorn, an editor of a major computer-industry magazine, *Datamation*, said several years ago that the analogy between computer programs and books is so extremely far-fetched that there is no justification for extending the law regarding copyright as applied to literary works to computer programs. His other main reason for not wanting to use the copyright mechanism for computer programs was that "eventually and probably sooner than most believe, it will be possible to protect a program by taking advantage of the computer hardware." What he meant, essentially, is that simply by programming the program properly, its secrets could be kept inviolate from would-be trespassers or despoilers. What he refers to are "execute-only, read-only" configurations within the computer for its self-protection.[30]

Nevertheless, there are likely to be some very interesting developments within this whole area before 1985. To know what is really going on, perhaps the interested individual or organization should keep an eye on CONTU—the National Commission on New Technological Uses of Copyrighted Works. Its Congressionally stated function is to recommend periodic revisions of the 1976 Copyright Law, which went into effect January 1, 1978. The NBS study by Saltman stated, in conclusion, "without copyright for computer-readable works, increased secrecy, cut-throat competition, and lowered opportunity for recognition of creative talents will result."[31] I agree.

# DOES IT PAY TO BE IGNORANT?*

In some dim dark year not so long ago—the 1940s, to be exact—there was first a radio show, then a TV program entitled, "It Pays to Be Ignorant." That odd statement, although likely to shock any educated person at first hearing, has a definite point to it. It *does* pay to be ignorant; it pays those who would exploit and use the unknowing for their own purposes. In a democracy, the first obligation of the participating citizen is to become aware of what is going on, within the necessary limits—for governmental affairs—of genuine national security considerations and—for the public sector—of the right to privacy for the individual. The federal government, one may be sure, will do all in its power to strive to have a monopoly of knowledge relating to military and foreign affairs; industry and business and finance, for reasons of profit, will attempt to keep to itself all facts pertaining to industry and business. But somewhere in all this, the rights of the individual are likely to be overlooked or even deliberately ignored.

With the coming of the widespread ability to store in a computer's data bank all sorts of information—not necessarily facts—about an individual, privacy is rapidly becoming an obsolescent, if not obsolete, word. As the situation now exists, computer data banks—somewhere—may have the following information about any individual: organization memberships and activity, properties owned, security clearances, FBI raw data, medical records, jobs held, unemployment information, credit ratings and detailed data, taxes paid (or unpaid), personality and IQ ratings, travel records, insurance facts, educational background, criminal records (or charges)—and those are only a few. Each one of the original occasions for making the records can probably be well-justified—but how much guarantee is there of the anonymity or confidentiality of the total material?

What about the role of the library in dealing with the seemingly irreconcilable dilemma of gathering and distributing information and also maintaining the individual's right to privacy? Communist Russia has libraries that keep permanent records of who has read what. More and more, as libraries turn to computer-based circulation management, such data lasts only a very short time. But there are still, of course, a very great preponderance of libraries that keep cards in each book, and these list the signature of the borrowers.

Only recently the new ultra-Right organization, the Moral Majority, tried to find out by court action exactly which schools in the state of Washington had borrowed a sex-education film from the Seattle Public Library. Fortunately, their efforts were suspended when a public hue-and-cry was raised. Out of this case just possibly might have come a definitive federal court ruling on the sanctity of library circulation records, still today a moot question.

Let's suppose—a not-*too*-unlikely hypothesis in the light of some recent Burger Court decisions—that the right to freedom of information is eventually held to be more vital—or even more constitutionally viable—than the right to privacy. There would seem to be two likely consequences: first, a greater effort to *maintain* privacy will be made by individuals, and second, as a sort of corrolary,

---

*Much of the material in this section first appeared in "From Books to Bytes," *Library Journal* (October 1981): 1877-80.

an even more pronounces tendency will develop to find out all possible data from every possible source about each person. In all of this, the librarian will be the man-in-the-middle, the one who will be asked to find out more, and who will himself or herself be forced to reveal more. It is not a pretty prospect.

The new technology that has come into the publishing and the library scene within the last few years brings with it concomitant problems and issues. If it is true, as a recent survey of "representative samples of librarians, publishers, and technologists" predicted, that electronic publishing of reference books (the so-called "electrobook") and journals will be the dominant style by the end of this decade,[32] then the effect on librarianship will be, must be appreciable. What sort of intellectual freedom problems will the "information-handler" (or whatever cognomen will be applied by the end of this century to what some of us still call "librarians") face? Perhaps the most obvious have been discussed above—those relating to the conflict of freedom and privacy. But there are others. F. Wilfrid Lancaster, Laura S. Drasgow, and Ellen B. Marks are highly optimistic that perhaps the greatest threat to intellectual freedom—lack of access—will not eventuate; in fact, their "scenario" for the year 2001 states, "fortunately, the electronic networks developed in the past twenty years have not created an information elite but have improved access to information for all segments of society."[33]

I beg, not so humbly, to differ. From my observations of the trend toward the so-called "paperless library," it seems to go hand in hand with a considerable drop in subscriptions to scholarly periodicals (which may very well result in the demise of quite a number of worthy journals, no matter how the Copyright Clearance Center and the projected National Periodicals Center likewise fare) and an astronomical rise in the costs of scholarly books—along with a drop in their sales. I am not so much in conflict with Lancaster, et al. on the possible expansion of the "information elite" to a wide-ranging information public as I am concerned with what seems a very likely continuing trade-off of machines for books. To my own knowledge, academic libraries that should have known better have traded off hours of service, book and serials purchases, even staff, for membership in information utilities. I think they—and their patrons—have lost by this exchange.

My own private utopia for the libraries of the next century results from neither questionnaires, nor Delphi studies, nor any other variety of social science heuristics. It is purely a hunch—an informed guess, if you will—based on rather a lengthy period of observation and participation in American librarianship. The Oboler Utopia—circa 2001 A.D.—will include the libraries of America as *not* linked by any national network, as *not* obliged to use VDT terminals instead of book stacks, as still maintained by librarians very similar to the librarians of the 1980s— some technology-minded, some management-minded, some user-oriented, some "bookmen," and even some a mixture of all of these.

The trend away from federal subvention and control of information will mean more local control of information and more opportunity for local access to it. There will, of course, still be censors and censorship—both within and without the library profession—but there will be a glorious opportunity for the 1980s "young bucks" among our profession to lead toward what may be a really utopian vision—the American library, free to do what it does best, provide "the right book for the right person at the right time."

# NOTES

[1] Alan Brier and Ian Robinson, *Computers and the Social Sciences* (New York: Columbia University Press, 1974), p. 278.

[2] Ibid., p. 277.

[3] Oceana Editorial Board, *Privacy–Its Legal Protection*, rev. ed., (Dobbs Ferry, NY: Oceana Publications, 1976), p. 101.

[4] Ibid., pp. 103-4.

[5] Richard H. Rovere, "The Invasion of Privacy (1): Technology and the Claims of Community," *American Scholar* (Autumn 1958), pp. 413-21; August Hecksher, "The Invasion of Privacy (2): The Reshaping of Privacy," *American Scholar* (Winter 1958-59), pp. 11-20; Granville Hicks, "The Invasion of Privacy (3): The Limits of Privacy," *American Scholar* (Spring 1959), pp. 185-93; Gerald W. Johnson, "The Invasion of Privacy (4): Laus Contemptionis," *American Scholar* (Autumn 1959), pp. 447-57.

[6] Rovere, p. 416.

[7] Ibid., p. 421.

[8] Hecksher, p. 13.

[9] Ibid., p. 20.

[10] George Meredith, "Modern Love," stanza XLIII; in *The Poems of George Meredith*, v. 1, ed. Phyllis B. Bartlett (New Haven: Yale University Press, 1978), p. 141.

[11] Hicks, p. 192.

[12] Ibid., p. 193.

[13] Johnson, p. 456.

[14] Quoted in Johnson, p. 451.

[15] Ibid.

[16] Rovere, p. 421.

[17] Alan F. Westin, "Life, Liberty, and the Pursuit of Privacy," in Donald P. Landa and Robert D. Ryan, eds., *Advancing Technology: Its Impact on Society* (New York: Wm. C. Brown Co., 1971), p. 361.

[18] Ibid., p. 363.

[19] Ibid., p. 372.

[20] Ibid.

[21] Most of what follows is based on Roy G. Saltman, *Computer Science & Technology: Copyright in Computer-Readable Works: Policy Impacts of Technological Change* [U.S. Department of Commerce, National Bureau of Standards. NBS Special Publication.] (Washington, DC: GPO, 1977.)

[22] Julius M. Marke, *Copyright and Intellectual Property* (New York: Fund for the Advancement of Education, 1967), pp. 92-93.

[23] Saltman, p. A-7.

[24] "How Silicon Spies Get Away with Copying," *Business Week* (April 21, 1980): 178-88.

[25] Saltman, p. A-11.

[26] Ibid., p. A-13.

[27] Ibid., p. A-14.

[28] Ibid., p. A-15.

[29] Ibid., p. A-16.

[30] Philip H. Dorn, "Programs Are Not Books," *Datamation* (Nov. 1977): 233.

[31] "NBS Study Recommends Copyright Changes to Protect Computer-Readable Works," *Journal of Library Automation* (March 1978): 74.

[32] F. Wilfrid Lancaster, ed., *The Role of the Library in an Electronic Society* (Urbana-Champaign: University of Illinois, Graduate School of Library Science, 1980).

[33] Ibid., p. 188.

# 7 Toward a 21st-CENTURY NATIONAL INFORMATION POLICY

We have three coincident revolutions in our civilization: the information explosion, the computer society, and the communications revolution. All are post-World War II events; each interacts with the others. The librarian is, *must* be, involved with all of these; the only question is whether he or she affects *them* or he or she is affected by their influence. Or, putting it another way, the librarian—an information and communications specialist since the profession of librarianship began—cannot hide behind a love of books and people and a loathing for the machine, and still accomplish the librarian's primary mission in the world today.

## THE MACHINE AND FREEDOM

The plethora of information and its ready availability via the wired society and telecommunications have encouraged doomsayers who want no part of the flood that threatens to engulf research on every level and on every topic. They are all like Ned Ludd, who, in early nineteenth-century England, solved what he saw as the otherwise insoluble problem of industrial mechanization by taking a very direct approach—he tried to break the machines likely to displace him. So some librarians would stem the information tide by not using the machines that carry it. To change the metaphor, they are bound in this instance to be ineffectual Canutes, since the trend toward providing more and more information and toward making as much of it as possible readily available is as seemingly irreversible as the tide.

There is another argument against Luddism. It is what has been described as "epistemological Luddism," considering "(1) the kinds of human dependency and regularized behavior centering upon specific apparatus, (2) the patterns of social activity that rationalized techniques imprint upon human relationships, and (3) the shapes given everyday life by the large-scale organized networks of technology."[1] His point, of course, is that we need to stand back and consider the true usefulness, the indispensability, or even partial or whole dispensability of the monstrous constructions we have set up for ourselves. The writer of the above suggestion, Langdon Winner, also suggests a few Luddite tactics that just might be useful in trying to appraise the value of technology to libraries and their users.

One simple one is to "dismantle or unplug a technological system in order to create the space and opportunity for learning." Another "is that groups and individuals could for a time, self-consciously and through advance agreement, extricate themselves from selected techniques and apparatus." Still another, he suggests, would be to "try disconnecting crucial links in the original system for a time and studying the results." And, finally, he proffers the almost obvious thought, that "the best experiments can be done simply by refusing to repair technological systems ... as they break down."[2]

Now, probably all of this sounds quite destructive, even childish, or futile. But there is a very serious side to all of these suggested ways of dealing in extremis with a self-built Information Monster. The devotees of the Great God Machine *do* have their counterparts in the library profession. All too many changes in methods of proffering traditional library services—or, more commonly, in unnecessary foisting of untraditional responsibilities on libraries—seem to have happened or have been planned simply because there was a machine-way of being able to do it. And, in reaction to the fanatics, the zealots, there are those in librarianship who would be almost as fanatical in their disdain, their disregard, for mechanical or computerized or telecommunications possibilities for the improvement of library service.

Let us step back a little in time. The megamachine of ancient Egypt was a combination of the scribe and the papyrus. One of the earliest available pieces of Egyptian writings says, "The scribe, he directeth every work that is in this land."[3] As Lewis Mumford says, "Action at distance, through scribes and swift messengers, was one of the identifying marks of the new megamachine...." The invention of writing linked far distances and disparate individuals and groups together; the printing and circulation of books, as Mumford has said, led to "the enrichment of the collective human mind...."[4] And the rise, in the seventeenth century in Europe, of the printed scientific paper was responsible for the nineteenth century industrial revolution.

What will the twentieth century combination of telecommunication and the computer cause? Mumford warns of "the peculiar fascination automated systems would have for autocratic minds, eager to confine human reactions to those that conform to the limited data they are capable of programming."[5] This is, of course, a distinct threat to intellectual freedom.

Mumford calls the computer "the ultimate 'decision-maker' and Divine King, in a transcendent, electronic form...." He describes its "authentic divine characteristics: omnipresence and invisibility."[6] But, as should be obvious, the "omnipresent" computer has its difficulties, its disabilities. The best-known, of course, is "GIGO"—garbage in, garbage out. The machine can only produce what is put into it. Secondly, the computer more or less molds the *kinds* of questions that are asked of it. And, perhaps most important, as Mumford says, "for all its fantastic rapidity of operation, its components remain incapable of making qualitative responses to constant organic changes."[7] In other words, those changes that are not subject to qualitative measurement or that cannot be observed objectively are really outside the scope of the computer.

Mumford cautions us that hurrying into the temple of megatechnic society is not always guaranteed to give good results. This counsel of caution is not intended to in any way impede *normal* progress, and *normal* recognition of the virtues of the machine on any level. But librarians must know what they want and what they are doing before going any further into this potential quagmire.

Vincent Barabla, director of the U.S. Census Bureau, has pointed out that "today we are moving into an era of utilization in which the emphasis is on the ultimate use of information, not on its production." In fact, he says, "these technological advances simply mean more information in an already crowded 'in' basket to sort through as we confront the problems we must solve." He has provided a very useful list of what the information providers must do in the critical years remaining in this century:

Understand the needs for information and the underlying reasons for those needs.

Understand the uses and limitations of information and be able to explain these strengths and weaknesses in terms the user can understand.

Involve users early and during every step of the knowledge-utilization process.

Insure broad awareness by establishing a forum for users and potential users of information.

Learn to communicate in the language of the user and develop an understanding of the constraints under which the user must operate.[8]

As for the information users, he suggests that they become part of a team with the information providers, plan for whatever information is needed, and, finally, "accept the fact that good information requires hard work to obtain and properly utilize." Facts alone, in other words, are not enough. This whole field of knowledge utilization and transfer has been coming to the fore in recent years. Knowledge utilization has been described as "a complex process involving political, organizational, socioeconomic, and attitudinal components in addition to the specific information or knowledge."[9] It has further been defined as "a process that converts raw data into intelligence needed for decision making, which may ultimately lead to action."[10]

How does all this affect the librarian of the 1980s, 1990s, and beyond? It makes his or her responsibility into much more than an intermediary responsibility. If he or she is to be "part of a team," then a broader knowledge than that afforded by a bachelor's degree in some narrow field plus a master's degree in library science is obviously essential. Continuing education on a broad level (a lifetime if you will, of continuing study) is a sine qua non of the new librarian.

Let us imagine that some agency created an interconnected computer services system that combined every terminal, information processor, and data bank available in the U.S. or the world. This would certainly, as has been suggested, give every user just about the maximum flexibility in using data-processing services. It would produce a sort of a computer utility, which would resemble the present Bell system, or even become part of it.[11] There are, of course, dangers in such a monopoly or semi-monopoly. Roger Noll suggests that "certain features of the communications computer industries lead to single 'ruinous' or 'unfair' competition that will result in monopoly or oligopoly even if economies of scale do not warrant such a development."[12] As is readily apparent, we are rapidly approaching the positing of what looks like a dilemma which cannot be solved. Herbert A. Simon has said, "the effectiveness of democratic institutions, as well as people's confidence in them, depends in part on the soundness of the decision-making processes they use. Democracy does not require a town meeting in which every member of the public can participate directly in every decision. But it is enhanced by an open and explicit decision process that enables members of the public to judge on what premises the decisions rest and whether the decisions are formed by the best available facts and theories. In fact, such a decision process is a precondition to intelligent public participation of any kind. To the extent that computers can contribute to its growth [the decision process that is open and explicit and informed], they can strengthen democratic institutions and help convent public feelings of helplessness and alienation."[13]

This seems to be quite an important issue. If we are simply plunged into the computer web, we are not going to accomplish as much nor feel as good about doing it as we would otherwise. The utmost possible participation by those who are going to do the work and those who are going to benefit from it is certainly to be encouraged.

We all have the right of access to truth. There are those who can uphold the cause in the halls of Congress, in state legislatures, in city councils, and on school and public library boards. The American, after all, began a several-hundred-year run on the stage of history by fighting restrictive authority—by Patrick Henry's desire for "liberty or death," continued by Justice Holmes' call for freedom for the right to express dangerous thoughts—and is even today well aware of the consequences of enforced ignorance.

Whatever one may say of the censor, surely he or she or it (if a faceless bureau assumes that role) never lacks for opposition. The act of promulgation and dissemination of information is very public; so is its other side, censorship. The willingness to face the mud-spattered obloquy of moral or political inquisitors—or even mechanically based ones!—is the price the exponent of intellectual freedom must pay. It is little enough to give for a cause that is of far more than casual or current importance. Every battle won against the censor is a step toward winning the eternal war for truth.

A Congressionally supported group, the Office of Technology Assessment, issued a report in 1981 as to their judgment of the impacts on society of national information systems. Although it was intended to be a popularized introduction to such systems and their effect on public policy today, it also served as a very good peep into the future "likely evolution of the computer and information industries."[14] They looked both at trends in the rapidly developing technology of computers and at what should be the contours, the gestalt, of the national information policy.

# PREDICTIONS AND REALITIES

They came to some interesting and valuable conclusions on both matters. As to the next decade's trends in computer technology development, extracted here are some of the more significant of the report's conclusions:

1. Computer electronics are experiencing an extraordinary drop in price, increase in power, and reduction in size.

2. There will be great expansion in the number of computers being used in business, education, and the home.

3. Computers will be used as components in a wide range of consumer products.

4. New products and services based on computer and telecommunication technology will become more available.[15]

In discussing the structure of the information industry (current and future), the report had the following conclusions:

1. Traditional information producers such as book publishers, newspapers, and network broadcasters will be converting their services into computer and telecommunication-based offerings.

2. New information services will be transmitted to the home over telephone, cable, and broadcast carriers. Some of these services may be integrated with in-house computer systems and video disk and tape units.

3. Libraries will extend their services beyond mere provision of books into offering computer-based services. Such new activities may conflict with the new commercial in-house services....[16]

In elaboration of this last reference, the report describes no less than nine categories that comprise the information industry. These include:

1. producers of primary information (books, journals, research studies, etc.);

2. producers of secondary information (indexes, bibliographies, databases, microforms, directories, etc.);

3. communication companies (broadcast, cable, switching, etc.);

4. information distributors, agents, or brokers (online service, sales representatives, dealers, etc.);

5. information transactors (banks, lending institutions, investment houses, etc.);

6. consultants or contract suppliers of information (designers, developers, etc.);

7. information retailers or outlets (on-demand services, search services, etc.);

8. equipment or supplies companies (computers, micrographics, text processing, graphic arts, etc.); and

9. popular media organizations (news, education, advertising, etc.).[17]

This is certainly a "skeleton list." And each of these categories includes a large number of types under that "etc." addendum. Just as one example, in the category of "information transactors" are those businesses whose normal activity generates information that may very well be of benefit beyond the individual firm. As the OTA report says, "more and more of this information is in computer-readable form and, furthermore, is possibly valuable to some other party."[18] The bank or airline or credit-card company that will hesitate to use the constantly growing volume and quality of valuable (to someone) information about their clients or customers is not likely to be among the "principal members of the information industry," as predicted by the OTA report for "data transactors," generally.

The OTA report faces up to three so-called "fundamental" values of information—all in likely conflict in the years to come. First, there is the tension between public and private values. Public values are shown by freedom of information laws, which are almost certain to conflict, as they say, "with individual or proprietary concerns (reflecting rights to privacy)." The computer has opened up less costly and time-consuming access to public records that have previously really been inaccessible because of the effort it took to get needed data from manually operated bookkeeping systems.

Then, there are commercial versus public values, as clearly evidenced in the information industry's desire to sell what public libraries give away. Another example of this, according to the OTA report, is "the competition between Government-collected data, made available through freedom of information laws, and commercial data services." Finally, in this regard, is the conflict between commercial and private values, most clearly shown in the widely known practice of "computerized mailing lists that may be compiled from third-party information sources without the knowledge or consent of the individuals involved."[19] (Anyone who ever subscribed to a national magazine or has given a donation to any special interest soon has a mailbox full of unsolicited "junk mail" as proof of how often and widely this happens.)

Perhaps the most important contribution of the OTA report is its clear, specific description of the conflicts and tensions incident to the changing role of information in our society, in the light of the opposing values with which it must deal. As the report states, "information technology is changing the relative importance of those values and throwing them into conflict." Since, as they say, "the traditional rules and procedures of the U.S. economic and legal systems are oriented toward the commercial exchange of tangible goods and services rather than of information," there are bound to be policy problems. Information has unique qualities, such as its reproducibility, very low cost of such reproduction, availability for nearly instantaneous transportation, normally very brief lifetime, and non-additive value. Such characteristics, the report stresses, "have created policy problems with respect to computer crime, copyright and patent laws, the flow of data between nations, and property tax laws."[20]

The second value of information, privacy, is discussed in detail in chapter 6, but the point should be made here that the private value of information is, to a certain extent, the rights to privacy of commercial organizations, as well as individuals. As for individuals, there should be no question about true freedom, including the freedom to be left alone—as long as this privacy does not deleteriously affect others.

There should also be no question about the traditional American point of view—evidenced in school, library, and museum support—that there is a public value to information, or, as the OTA report puts it, "the public interest in a free flow of information." Among the OTA report evidences of this are these: "a tradition of academic freedom and a system of open scholarly publication to promote the exchange of ideas and the advance of research; First Amendment guarantees, which ensure an unfettered discussion of political ideas, freedom of religion, and a free press; and freedom of information laws asserting the rights of citizens to know as much as possible about the actions of their Government."[21]

The OTA report, described herein before, is only one in a whole series of similar reports—governmentally and privately sponsored—attempting to give the parameters of a useful national information policy for the years to come.

One of the most incisive of these is the report titled *Telecommunications & Society, 1976-1991*, by Arthur D. Little, Inc. (1976) described earlier. Of the two factors that go into "compunications," this study deals mostly with communications, rather than computers. It points out that "telecommunications is so close to the heart of our economic infrastructure that changes in its technology can produce impacts of great extent and variety.... Future broadband capabilities [offer] possibilities for human interaction with one another and with basic data banks, libraries, and other information resources that go far beyond any of the concepts that have led us to our present telecommunications position."[22]

As to the future of the telecommunications industry, the Little report sees the following:

> the telephone companies strong and dominant in the local switching, local loop, and business and household distribution system for point-to-point and even some broadcasting-type services;

> a wide variety of organizations, including AT&T, becoming active both in transmission and in value-added services, utilizing a variety of satellite and terrestial transmission systems and highly sophisticated computer switching and control techniques and the like to compete with one another for long distance transmission operations;

> possible eventual opening up of the frequency spectrum to greater utilization by individuals and organizations operating in a less structured fashion than is now necessary;

> possible evolution of a new government-sponsored satellite carrier in competition with AT&T and other commercial common carriers in the long distance point-to-point and broadcast satellite markets; and

rebundling of functions for a number of existing telecommunications industry components and other industries, leaving most of these with only part of the activities they now perform.[23]

Their perspicacity in 1976 is evidenced by how many of these prophecies, in whole or in part, have already been fulfilled, a few years later. They look *very* likely to be correct in their further statement that "the telephone companies with their earlier position in the home distribution system field, their much greater existing coverage, and their far greater financial and technical resources—will be the winner in any contest to perform the basic broadband household connection job."[24]

They even foresee what they call "the (at least partial) potential replacement of other distribution mechanisms (such as mail) and resources (such as libraries) by electronic techniques...."[25] However, here I sharply differ for there are many other functions and services already and potentially available from the U.S. Post Office and from American libraries that compunications cannot replace—but only supplement.

Of the five scenarios offered by the Little report for the future of American telecommunications—the obsolence and eventual demise of the present U.S. Postal Service (to be replaced by an all-electronic service); the development of Citizens Band (CB) radio into national, all-citizen intercommunication; broad band distribution of broadcasting to the home; the growth of a public service satellite system; and AT&T's taking over the broad band communication networks—only one might be of real interest to the library world. M. F. Roetter, the author of this particular chapter in the Little report, foresees (for 1984) the following as a possible complication arising from the last scenario, as follows:

> ...the operation of the interlibrary wide band communications network in the Boston area ... was being delayed by a court order obtained by a group representing copyright owners, who were demanding that payment be made when material was 'borrowed' via telecommunications. The network itself was based upon a mixture of local (on-premises) wide band coaxial and optical fiber systems, linked by millimeter wave radios. The principle that fees should be paid to copyright holders had been established some time before ... by ... Congress.... The argument was about the methods of collecting and disbursing payments, as well as the amounts involved. In particular, the copyright holders wanted detailed records of all copying and lending transactions, whereas the libraries argued for a flat-rate schedule of payments.[26]

The passage concludes, somewhat pessimistically, by saying that it took a year and a half to resolve this controversy, "during which time the financial position of some libraries became so acute they were forced to close four days a week."[27]

The Little report does *not* predict that any one of these "scenarios" will really become fact; rather, it calls them "a sample of possibilities for the future, derived from considering interactions between telecommunications innovation and our social environment."[28] They place equal stress on both, in their "summation." Their awareness of the fantastic potentialities of technologic changes for the 1980s make this one of the most valuable documents available for the "futurist" in the field of telecommunications.

Since several of the "scenarios" predict the at least partial "potential" replacement of such resource-suppliers as libraries by electronic techniques, what they have to say about the potential *denial* of telecommunications is of great importance. They stress that "far more than other activities (such as transportation)..., use of telecommunication systems directly involves matters such as security, privacy, and the right to exercise free speech ... the extension of means available for communicating makes the issues more complex; one man's 'free speech' can interfere with another's, for example...."[29]

All this brings to the fore the importance of governmental regulation and subsidy. The report says that "the regulators ... often make the basic conditions as to which organization will survive and prosper and which will decay."[30] One is reminded of the battles of the American railroad giants in the latter half of the nineteenth century; the amount of subsidy and type of regulation decided which of these high-risk ventures lived or died. They conclude by arguing strongly against the establishment of a Big Brother-like single government monopoly of telecommunications. They favor a mix of government subsidies, specialized government-supported agencies, private regulated monopolies, and lightly regulated free market competition. Their final word (which is, perhaps, obvious) is, "there is no ideal or optimum solution to these problems, so they will continue to be with us."[31]

Another Arthur D. Little, Inc. report germane to the topic of this chapter is one prepared for the National Science Foundation and published by the American Library Association in 1978. Its title, *Into the Information Age: A Perspective for Federal Action* reflects the period—somewhat euphoric—during which it was prepared. The Reagan era political climate, compared to that Carter period, makes it very unlikely that the recommendations of this report will soon—if ever—become more than just that—recommendations. The writers saw their task as trying to find out "how to make information better for our society."[32] Within the broad scope of "information," their major concern was scientific and technical information (acronymed as STI). But, they said, "*Scientific and Technical Information (STI) needs to be augmented with Societal Information (to make STSI) to be most useful for problem-solving.*"[33]

Even though, as they said "*the existing STI infrastructure in the U.S.*, that is, today's complex of scientific and technical journals, data banks, indexing and abstracting services, repositories and libraries is without doubt the best in the world," there are problems. For example, "some of the traditional STI systems—e.g., libraries, journals, abstracting and indexing services—are those most requiring overhauls to meet the current problem-solving needs."[34]

What is of most significance to American libraries and librarians in this so-far hardly tapped resource-report are its specific recommendations and predictions for libraries. First of all, they see the American *public* library as, at best, obsolescent. They assert that "short of a basic change in the role and perception of the public library, there is every reason to believe that it will continue to decline in relevancy, at an accelerating rate."[35] For public libraries they recommend that they "*begin a 'second tier' of services that are for-fee, using market mechanisms on a partially subsidized not-for-profit basis (subsidy of investment, not operations.)*"[36]

Academic libraries were characterized by the report as being in as bad a financial bind as public libraries; it called for them to "move away from their traditional functions as custodians of materials and more toward creation of newer (electronic and/or microform) campus information dissemination systems, and toward an increasing role of serving as information intermediaries." They add that "it is unclear now who will own, operate, or control the electronic information delivery systems which will serve colleges and universities ten to twenty years from now—publishers, state governments, university systems, commercial firms, professional associations, the federal government and/or local librarians."[37] Actually, the potential suppliers go far beyond these, as indicated in an earlier Little report, described heretofore.

Incidentally, the report discussed in the last few pages was branded as obsolete by one of the most perceptive thinkers in this area, Richard DeGennaro. DeGennaro's basis for this conclusion was what he described as "a massive shift in government spending policy."[38] The ups and downs of federal spending policy, in this writer's opinion, do not lend themselves to absolute, sweeping judgments. Just as the Reaganites preach economy and deprecate non-defense spending, so a new regime (congressional and presidential) can come up within a very few years and find the Little reports fresh and inspiring.

What really matters are not the day-to-day, even year-to-year vicissitudes of the relationship between the federal government and the Information Age. What will ultimately decide the modus operandi for information seekers of the decades to come will be what best satisfies them. I agree with DeGennaro that the market place is the great decision-maker; whether the pattern solidifies in one shape or another it is not possible to foretell with any degree of certainty. What is absolutely certain is that there will be a pattern, and that libraries will be a part of it in some form.

Somewhat earlier, a panel on Science and Technology met with the U.S. House of Representatives Committee on Science and Astronautics to discuss the management of information and knowledge. After a two-day meeting, they emerged with a report that surveyed in detail the current and likely effect on our society of the unexpectedly rapid development of computerization, as well as the mid-century "revolution in communications technology." McGeorge Bundy, then president of the Ford Foundation, stressed the significance of "the intersection of important questions of intellectual freedom with forms of educational finance," by which he was referring to the "distortions and dangers of channeling disproportionate amounts of Federal aid for graduate training, research, and institutional development via the defense budget."[39] He warned that "in the era of information explosion, societies can become paralyzed over a plethora of facts and the absence of obvious conclusion. Or they may freeze," he said, "when the indisputable facts and necessities offend received values and conventional wisdom."[40]

The retired Chief Justice of the United States, Earl Warren, in a keynote address, said that, in dealing with the rapid movement of arts, science, and technology, "the job of every generation is to find an accommodation for them which will not dehumanize us or distort the ideals we have long held but not achieved for American life."[41] He also stressed the need for restraining and regulating the use of electronic equipment to invade privacy, saying that "the unwarranted use of ... uncritical information often irreparably harms the standing of the affected individual in the community. It also," he said, "creates dissension between individuals and groups and makes for disunity."[42]

Daniel Bell, well-known Harvard sociology professor, told the group that since America is clearly the first "postindustrial" society (in which most of the labor force is involved in services), it is "organized around information and utilization of information in complex systems, and the use of that information as a way of guiding the society." He sees the "elements of interdependency and complexity in a postindustrial society" as something "genuinely novel in history."[43] In the U.S. from now on, the innovative sources are "the intellectual institutions, the universities, the research institutes, the research corporations." (And, may I add parenthetically, this makes the library even more important to the inevitable organization of change in our society than in the preceding two centuries of first agricultural, then industrial American society).

The well-known "futurologist," the late Herman Kahn, formerly director of the Hudson Institute, echoed Daniel Bell's stress on the post-industrial society. In looking over the last third of the twentieth century, he saw "a relatively multipolar, relatively orderly, relatively unified world ... [with] enormous growth in world trade, communications, and travel...." He also saw "a basic, long-term, multifold trend toward ... accumulation of scientific and technological knowledge ... [and its] diffusion...."[44]

The development director of the International Publishing Corporation, Stafford Beer, told the panel that "technology now seems to be leading humanity by the nose."[45] Further, he claims that "we are not going to do anything at all about the management of information and knowledge towards the regulation of society as long as we think in data-processing terms." He calls for "the machinery to transform data into information, and the machinery by which that information may be used to innervate society." He sees the basic problem of our day as "informational overload." In fact, he tells us, "the private citizen seeking knowledge is inundated by information which is virtually free."[46]

So, continues Beer, to manage information properly we must refine and filter a "massive" overload of it. Curiously (and he was not speaking directly of libraries and their potentialities here), he says that "we might well say that it is a problem not so much of data acquisition as of right storage; not so much of storage as of fast retrieval; not so much of retrieval as of proper selection; not so much of selection as of identifying wants; not so much of knowing wants as of recognizing needs...."[47] And doesn't all that sound a great deal like the American Library Association's early twentieth-century formula for the American public library, "The right book for the right person at the right time"?

The director of Stanford University's Computation Center, Paul Armer, warns of the "actually accelerating" rate of change—particularly in computers and communications, and calls for a universal "continuing education" program to keep us all up with technology.[48] Osma A. Wiio, a Helsinki University business professor, speaking on technology, mass communication, and values stresses possible effects of computer-technology progress, especially "the loss from individuality and the dangers of computer data systems for the individual."[49]

The dean of the University of Texas' business college, George Kozmetsky, sees such technologic devices as computers as "natural extensions for the individuality of teaching as well as for the individual's development of creativity and inventiveness."[50] For all the "non-routine pursuits" necessary to cope with the rest of this century, he sees the need for "large quanta of technical and intellectual resources such as ... scientists..., engineers, and other professionals and service personnel and technicians as aides to the professionals." He also sees the need for "relevant and up-to-date information" to help solve the "non-routine" problems.[51] He optimistically claims that "the current trend of computer application to bibliographical search microfilming, microfilm cards, print reading, linguistics and library sciences does help narrow the transference of technical data gap."[52] Or, in other words, computers will speed up necessary research, he thinks.

In his "Summary Views and Comments" at the end of this report, lawyer L. Harvey Poe, Jr., called on us to look at technology "not ... as a threat to our freedom and integrity as men, ... but as a potent implement, which, if transcended and bent to our purpose, will make us almost more than men." He said that we "must and *can* restrain [the use of the new technology to invade the privacy of others] ... through law; perhaps this should be an addition to our Bill of Rights...."[53]

This entire report seemed oriented toward the urgent need for preserving our humanistic values, while using the new information technology for the material advancement of man.

# NOTES

[1] Langdon Winner, *Autonomous Technology: Technics-Out-of-Control As a Theme in Political Thought* (Cambridge, MA: MIT Press, 1977), p. 331.

[2] Ibid., pp. 332-33.

[3] Lewis Mumford, *The Myth of the Machine: Technics and Human Development* (New York: Harcourt, Brace, 1967), p. 192.

[4] Ibid., p. 285.

[5] Lewis Mumford, *The Myth of the Machine: The Pentagon of Power* (New York: Harcourt, Brace, Jovanovich, 1970), p. 189.

[6] Ibid., p. 273.

[7] Ibid.

[8] Vincent P. Barabla, "The Future Role of Information in American Life," in Norman Cousins, ed., *Reflections of America: Commemorating the Statistical Abstract Centennial* (Washington, DC: U.S. Bureau of the Census, 1980), p. 196.

[9] Judith K. Larson, "Knowledge Utilization: What Is It?" *Knowledge* (March 1980): 424.

[10] Barabla, p. 196.

[11] Roger G. Noll, "Regulation and Computer Services," in Dertouzos and Moses, eds., *The Computer Age* (Cambridge, MA: MIT Press, 1979), p. 256.

[12] Ibid., p. 255.

[13] Herbert A. Simon, "The Consequences of Computers for Centralization and Decentralization," in Dertouzos and Moses, eds., *The Computer Age*, p. 226.

[14] *Computer-Based National Information Systems: Technology and Public Power Issues* (Washington, DC: GPO, 1981), p. iii.

[15] Ibid., pp. 123-43.

[16] Ibid., p. 165.

[17] Ibid., pp. 163-65.

[18] Ibid., p. 165.

[19] Ibid., pp. 15-16.

[20] Ibid., p. 49.

[21] Ibid., p. 50.

[22] U.S. Dept. of Commerce, National Technical Information Service, *Telecommunications & Society, 1976-1991* (Springfield, VA: NTIS, 1980 [dated June 22, 1976] ), pp. 136-37.

[23] Ibid., pp. 139-40.

[24] Ibid., p. 141.

[25] Ibid., p. 142.

[26] Ibid., pp. 108-9.

[27] Ibid., p. 109.

[28] Ibid., p. 135.

[29] Ibid., p. 137.

[30] Ibid., p. 138.

[31] Ibid., p. 147.

[32] Arthur D. Little, Inc., *Into the Information Age: A Perspective for Federal Action* (Chicago: American Library Association, 1978), p. iii.

[33] Ibid., p. 5.

[34] Ibid., p. 7.

[35] Ibid., p. 115.

[36] Ibid., p. 117.

[37] Ibid., p. 125.

[38] Richard DeGennaro, "Libraries, Technology, and the Information Marketplace," *Library Journal* (June 1, 1982): 1047.

[39] U.S. House of Representatives, Committee on Science and Astronautics, *The Management of Information and Knowledge* (Washington, DC: GPO, 1970), p. 7.

[40] Ibid., p. 8.

[41] Ibid., p. 10.

[42] Ibid., p. 11.

[43] Ibid., pp. 14-15.

[44] Ibid., pp. 21-26.

[45] Ibid., p. 43.

[46] Ibid., pp. 45-46.

[47] Ibid., p. 46.

[48] Ibid., p. 82.

[49] Ibid., pp. 85-88.

[50] Ibid., p. 93.

[51] Ibid., p. 97.

[52] Ibid., p. 105.

[53] Ibid., pp. 127-30.

# AFTERWORD
## The Humanist and the Computer

A decade ago the computerization committee of the Japan Information Development Association issued a report, entitled *Information Society in Perspective*, which, among other things, predicted a "computopolis" as a likely possibility for the not too distant future. In this "computopolis," they foretold, there will be: "home terminals; computerized vehicle service; automated supermarkets; servo-ventilation; and local health admistration systems," as well as computerized transport, material goods-flow, and local government administrative systems—all related to computers.

One of the most basic results of computerization of our lives, they warned, would be excessive conurbation; that is to say, the mix among rural, small-town, suburban, and big-city life will cease to exist. Instead, "computopolis" would take over. This would affect—for the worse—family, small neighborhood, and basic community relations. Indeed, they saw the coming of "a sort of re-decentralization of brains and talents and re-decentralization of the opportunities for participation." They saw other deleterious results, humanistically speaking. For example, "meritocracy (would) dominate the educational process and aggravate ... social alienation in school." And although computerization of the city—which they called "informationalization"—might, paradoxically, foster creativity, along with added leisure, it would also be likely to bring added bureaucratic control, loss of privacy, and even added unemployment and crime.[1]

### THE INDIVIDUAL IN
### A COMPUTERIZED WORLD

The computer and all its byproducts are, of course, entirely dependent on what society does with them. The over-computerized public library has been classically described by Dan Lacy (in pre-silicon chip days) as follows: "mechanized, electronic libraries in which queries are translated into binary symbols that flash through batteries of tubes and seize on the relevant data from millions of volumes, all coded on magnetic wire, and assemble the desired information whipped out on an IBM printer."[2] Lacy, writing in 1956, predicted that "fifty years hence, as fifty years ago, the book as a physical artifact will not differ greatly from what it is today."[3] More important than conjecture is the not-too-difficult-to-perceive ineluctable *necessity* of this being so. The book *must* survive, as Lacy points out, because, "the vitality and freedom of books are ... essential to society's capacity for social invention and for that swift and purposeful change its survival demands."[4]

But books are not important only for their share in social, scientific, and technological progress. Lacy reminds us that "the defeat or the victory of human values goes hand in hand with the extinction or the flowering of books." In the world of mass communications, of "compunications" and "computopolises," books—not computers—will provide "a means for men to withdraw and search alone for the continuities of value and insight that give form and meaning to our human adventure. And this, too, means books."[5]

It is not, of course, true that the computer will necessarily bring less freedom, less freedom of the mind, unless it is used to do so. It is only a tool, and a tool has no values, no social instincts. But, as has been said, "new technologies—especially those involving electronics and computing—have made currently possible many forms of intrusion and control that figure as science-fiction fantasy in *1984*."[6]

The key to what the new technology of information can do, may do, is likely to do, is the control of such technology. When the modern computer first emerged, it was considered likely to be under the control of those who produced the hardware (that is, IBM, Sony, and similar companies). But the rapid, constant merging of the communications and computer-based operations (compunications) have brought, worldwide, the various telephone and postal and telegraph administration (commonly called PTTS) into the picture. Bell Telephone has been judicially freed to get into the computer business, while IBM today, for instance, "is not just concerned with computers, but is involved in defense telecommunications systems, service bureaus, copiers, printers, and business forms." Xerox, according to recent information, "has a goodly piece of the computer printing equipment, film, facsimile, mailing equipment, software, and time-sharing business"—as well as having recently bought Western Union International. RCA, in addition to its well-known involvement in radio and TV, also is concerned financially with "satellites, telephones, ... books and records, satellite equipment, antennas, ... videotape and disc equipment, and defense telecommunications systems."[7]

Various book, newspaper, and periodical companies have widened their horizons and gone into the information business on many levels. Some financial institutions "are offering time sharing, data bases," etc.[8] Even EXXON is acquiring small businesses in the information area and selling electronics equipment.

Who regulates these activities? Mainly, the Federal Communications Commission does. But, in various ways, so do the Interstate Commerce Commission, the Civil Aeronatics Board, the Postal Rate Commission, the Department of Justice, the Federal Trade Commission, and so on and on. As the Ganleys say, "the same sort of confused regulating jurisdictions, only worse, exist around the world...."[9]

In a brief but important essay, "The Anatomy of Modern Technology," Bruce Hannay and Robert E. McGinn have presented four "important technology-engendered aspects of the character of modern Western life...." These four, they say, are "scale, multiplicity, flow, and transience." All four apply to the effects of communications technology.[10] The fantastic number of applications of the computer and of telecommunications, their spread to global and even interplanetary dimensions, and their effect on the power of the individual are all indications of growth in scale and multiplicity.

What is perhaps less apparent is the pressure on Western civilization of the growing necessity for (to use Hannay and McGinn's words), "a greatly increased flow or 'throughput'—material (people, products, energy, and material resources), ideational (ideas, knowledge, and information), and experiential—in various spheres of activity." They point out that modern technology has both "made possible and mandated" increased flow through the various channels for distribution in society—"modern mass production, transportation, and communication technologies." Perhaps the most thought-provoking of this list of "technology-engendered aspects" is the last—transience.[11] The transitory, very short-term nature of a great many of the communications gadgets developed in the past few years is all too well known. In computer-technology, vacuum tubes were replaced by transistors, which were replaced by silicon chips, which will be replaced by—who knows? The fact is, say Hannay and McGinn, "many people sense they pay a psychic price by being suspended in a turbulent, flux-charged cultural environment."[12] Walk into any video-game arcade or tune in to any "Top Forty" radio station, and judge for yourself.

This brings us full circle to the most important link in the whole telecommunications/computer chain—the individual. Robert Louis Stevenson long ago pictured for us "the case of the poor young animal, Man, turned loose into this roaring world, herded by robustious guardians, taken with the panic before he has wit enough to apprehend its cause, and soon flying with all his heels in the van of the general stampede...."[13] And RLS's idea of "this roaring world" was nowhere near the facts of this century's information-hungry planet, crisscrossed with messages and news and facts and opinions, over wires and via satellites in space, imprinted on silicon chips and thrust through laser beams. This "poor young animal, Man" has much to do to ensure that he maintains his individuality, his raison d'etre in a world and an era where the individual seems to matter less and less. The one thing of which the Information Age has made us all aware is that the One is going to have a mighty struggle with the Many—the Id with the Mass. Who wins will determine the freedom of the mind for a long time to come.

## THE LIBRARIAN AS MEDIATOR

If the librarian is "the mediator in communication exchange," as Lester Asheim has recently described him or her,[14] then the librarian must accept the responsibility of performing that mediation. With the coming of compunications, the librarian has an even more vital duty in a mediatory role; we must face up to the double responsibility Asheim defines: "one, as the builder of the store of information; the other, as the intermediary between that store and the user's present need."[15]

There is no longer any possibility of even the Library of Congress' or the British Museum's holding all available information. But there is not only the possibility but the necessity of the librarian's acting in Ortega y Gasset's term, "as ... a filter interposed between man and the torrent of books."[16] If the librarian takes unprofessional advantage of his or her position to let prejudices control that filtering process, then that person is not a librarian, but a propagandist.

Asheim has offered a possible alternative metaphor to Ortega y Gasset's "filter" image, necessitated by the fact, as he sees it, that "the mechanical devices we now have for making our resources greater and more complex push us towards entropy unless we can devise some means for trimming the transaction itself down to human scale."[17] Any database user who has asked, say, MEDLARS for available information on such an unlimited, uncircumscribed topic as "cancer" or "stroke" knows that the machine-handlers will not even attempt to fulfill the request. It must be handled, as Asheim suggests. That is to say, "control of the flow is adjusted, through a variety of techniques and devices, to provide the pertinent information in the appropriate amount to be useful in the patron's evaluation, assimilation, and utilization of the information."[18]

If the danger of an overload of information, if the kind of stasis or entropy, in Asheim's term, is really so great, what can the librarian do beyond standing like some kind of Cerberus at the gates to knowledge? This attitude seems so contrary to library tradition that it is hard to recognize it as an ineluctible necessity in the current stage of the Information Era. Here again Asheim's views may be of assistance. He stresses that since "more than any other gatekeeper in the field of formal communication, the librarian is devoted to the tradition of individual service and individual response," then it is up to the librarian to *use* that tradition, carry out that function. Out of this will come, Asheim assures us, what he calls "traffic control in the information search...." And, mirabile dictu!, this action will, he continues, "not seek to inhibit the expression of ideas, but rather to facilitate their assimilable movement and display."[19] It is as if one were to attempt to cross a street full of speeding vehicles without taking into consideration their directions and speeds relative to the unwary pedestrian.

Asheim puts this in a simpler context, that of the librarian's longtime service tradition. As he says, "we have been dedicated to serving the user; we have fought long and hard for users' rights to freedom of access to ideas; we have conscientiously attempted to keep our own preferences from taking over in the selection process; we have attempted to facilitate the individual's own search rather than to construct a mystique around our procedures that would require their dependence upon our intervention."[20]

And all of this is made more essential by the technologization of much traditional library service—from interlibrary loan to reference service. Asheim's insistence on "sensitivity to human feedback during the transaction"[21] is eminently justified. It is the human, not the machine, who is, after all, basic to library service in the past, the present, and the foreseeable future.

Eric Hoffer, that self-appointed apostle to the common man, recently wrote, "when you automate an industry, you modernize a life or you primiticize it."[22] This, of course, applies to the working life of a librarian. Blind subservience to the machine—whether it is a computer, a telecommunicator, or whatever—is not likely to be in the best interests of either the machine-slave or the patrons served. It is only when the librarian, fully aware of the human implications, causes, and consequences of such work, considers and treats the tool *as* a tool that the needed work will be done humanistically, not mechanically.

Colin Norman, in considering what he called "the Knowledge Business"[23] in his book subtitled *Science and Technology in the Eighties*, said that in the competition between humanism and science, science/technology have so far been far and away the winners. As one indication, "government fund-of other cultural pursuits amounts to only a tiny fraction of that devoted to basic research."[24] There are other evidences of the extent to which "the god that limps" (the Greek god of fire and metalworking, Hephaestus, in Norman's metaphor) has affected the human side of work as related to technology. Norman cites Robert Howard, a writer on the effects of major changes in the telephone industry on the workforce: "experiences that telephone workers described ... [include] traditional skills made obsolete by new technology; jobs fragmented and downgraded to lower pay; work reorganized and rigidly centralized; workers subjected to automatic pacing and monitoring, oversupervision, and job-induced psychological stress."[25] All this may seem very familiar to last-generation-trained librarians working under current automating conditions.

It is surely not just coincidental that articles appear in library literature dealing with the results of the "job-induced psychological stress" to which Howard refers.[26] Even though Rudolph Bold's article only refers very briefly to "computer charging system failure" as one possible or likely cause of "librarian burn-out," the implication is there. Even though, admittedly, "stress would seem to be common among professionals," as Bold states,[27] it would seem to be at least prudent for everyone to take appropriate measures to keep stress at as low a level as possible.

There is little question that (referring to "the microelectronics revolution), as Colin Norman puts it, "it is in the workplace that the new technology will ultimately have its chief social impact.... No technology in history has had such a broad range of applications in the workplace." And the library workplace shares in these "broad ... applications ..." enough already (with strong prospects of much more to come) to have changed, in many respects, both formal library education and informal library inservice training.[28]

With this kind of milieu, how will intellectual freedom fare in the automated (or, at least, semi-automated) library of the future? Even a clouded crystal ball will reveal some onimous potentialities. Over-concern for the good of the machine may or may not mean under-concern for the human-being—but both library patrons and practioners need to take a good hard look at how fast they are heading and where.

No one is trying to turn the clock back to quill-pens and hand-stamping at circulation desks. But it is to be hoped that no one is unaware of the deleterious possibilities in over-reliance on machine read-outs and the machine-processing. The Asheim admonition, that "the mechanical devices we now have for making our resources greater and more complex push us toward entropy...,"[29] is very apropos.

The continuing task, the professional responsibility, of the librarian is to bring information and user together in the most accommodating, least expensive, and most freedom-of-information-promoting way.

## BOOKS AND BYTES*

A little over half a millenium has passed since Gutenberg brought Western civilization its most considerable technological achievement since the wheel—printing. And there has also passed a little more than a third of a century since I began my career as one whose adult life has been involved with the media. That last period—infinitesimal in eternity—has involved a literal revolution in librarianship as a profession, as a career, and as a service to mankind.

In the 1940s, the furthest reaches of library education involved what seemed then to be a truly revolutionary concept—to consider the film as a respectable addition to the almost entirely print-centered library of that day. Indeed, the film was seriously considered as a total replacement for the obviously soon-to-be-made-obsolescent (if not obsolete) codex book. The film, after all, had potential that no book or magazine or pamphlet could possibly match—it provided an actual sound and view of what print required imagination to interpret.

In a very few years, the orthicon tube brought into the home everything for which the film required a theater. Once again, the book was presumed to be on its last legs. Actually, the post-World War II era complicated the print-bound library's activities with still another new phenomenon—the paperbound book. Although Frank Schick predicted in 1959 that it would more or less take over, we are still living with both hard and soft-bound formats—although as binding prices skyrocket, we may yet follow the common European formula of first publication for all books as paperbacks.

These are but minor developments in the constantly evolving arena of information/communication. A much greater change was taking place in the library world—a new determination to implement one of the basic principles of all America—let alone the library. The national disgrace of differential treatment of the races in libraries, and in the American Library Association itself, led to a several-year dispute in and out of ALA conference meetings, climaxed by my revelation[30] that only ALA, among national professional-type associations, had separate and unequal chapters. Four southern states in the early 1960s still barred black members from state library association conferences and constituted "whites-only" chapters as sending councillors to ALA Council. By 1966, under threat of permanent expulsion from ALA, all were integrated, with *single* chapters and equal treatment in all state associations for all races.

But other areas in the never-ending struggle for intellectual freedom offered me even greater satisfaction through the years. From a small beginning, a decade and a half ago, in getting ALA to do more than publicize library censorship battles, there is now a whole array of ALA-connected agencies and groups, from the Office for Intellectual Freedom—whose establishment was a direct result of the ACONDA priorities for ALA set up at the memorable 1969 Atlantic City conference—to the Freedom to Read Foundation to John Philip Immroth's Intellectual Freedom Round Table to the continuing policy-setting work of the Intellectual Freedom

---

*Much of the material in this section first appeared in "From Books to Bytes," *Library Journal* (October 1981): 1877-80.

Committee. My own small part in all of these reflects my feeling that social responsibility (as it directly relates to libraries and librarians through fighting all kinds of censorship) is the highest priority of my profession.

Just as I came back from war service, academic librarians were envisaging what they felt to be a real change in their objectives. As Charles H. Brown wrote (as chairman of ACRL's Committee on Wartime Activities), "the attention has been shifting from studies of library organization and routine to the consideration of needs and definite methods of satisfying such needs."[31] This trend has, of course, continued until comparatively recently. Then the era of *management* suddenly arrived, with its panoply of assorted acronyms, from AAB to ZBB, with PPBS prominent along the way.

Fortunately, there is another counter-trend, which may well at least bridle the rampant management monster and put it in its appropriate place. This trend—toward giving first consideration to the *user* of the library—is flourishing in all types of libraries. The kind of detailed, individual concern evidenced in most special libraries is, of course, impossible in the large public or academic or even school library; but it is a good thing that the current attention paid to the user makes Charles Brown's 1946 dictum look good in 1981.

The academic librarian of today must, of course, first and foremost, deal with the exigencies with which all academe is dealing—lowered budgets, increased costs, willy-nilly modernization. Books keep getting published in the multi-thousands each year; new, exciting, useful periodicals start up constantly; changing curricula demand ever-increasing bolstering with all manner of media. The changes incident to AACR 2, the evolving national information policy and legislation, the inevitable development of home and office computer terminals—all these and many more make it even more important that the academic librarian resist the temptation to become more of a businessman than a librarian in the best humanistic sense.

The neophyte emerging from typical contemporary college training is not likely to have anything like the liberal-arts-based, general education that was prescribed in the 1930s and 1940s as the best prelude to the library science core curriculum. When not one but several accredited library schools have substituted COBOL and FORTRAN for the once-customary "foreign language" requirement, one can hardly expect the prospective librarian to insist on studying French or German or Spanish. And Latin or Greek? Forget them! They, of course, are not (key *sprechwort* of the educationists) "relevant."

This is not to gainsay the cleverness, the "with it" abilities of today's librarians. The young ones all seem to know a modem from CODEN, an acoustical coupler from a floppy disk, or a bit from a byte. They are fluent in what is the argot of the day; one hopes that they are equally as conversant with the Great Dialogue of the immortals. A little more Horace and Longinus and Boileau might even help in literary criticism of some of the current TABA winners. Ecological considerations are basically no different today than in Virgil's *Georgics* or Columella's *Res Rustica*. The economics of Say and Smith and Bagehot could be of help in confuting or confirming Friedman or Samuelson or Laffer. The sociology of a Comte or a Durkheim could help explain a Parsons or a Mead. Today's reference librarians all too often find skill at operating a computer terminal keyboard vastly more useful than a wide-ranging personal knowledge of where to look for what. The reference interview is beginning to be less significant than machine-handling dexterity.

Plato's *Republic* envisions the Philosopher-King as the hope of organized mankind. With the rapid and accelerating pace of development of information, creation, transfer, and storage, there is an absolutely reasonable vision that we who live by and with and for the selection and dissemination of knowledge can see as more than a utopian dream—the Librarian as Philosopher-Librarian. If it is true that whoever controls information in the Age of Information just ahead of us will truly control the world, perhaps the Librarian as Philosopher first and Librarian second will assume a dual role which is both inevitable and necessary. I see my generation as a transition group between the Librarian as Servant and the Librarian as Leader; I think it has fulfilled that function well. Fortunately for us, we have two amazing prototypes—one still a hale and hearty nonagenarian, Keyes Metcalf, and the other just recently gone, after more than a century, Louis Round Wilson. They set a high standard, to which the library generation emerging in the 1980s can well aspire. Philosopher-Librarians both—and surely the kind of leaders librarianship will always need.

Henry Steele Commager has pinpointed the major difference between the founders of our country, those whom he calls "the American Philosophes," and their eighteenth-century European counterparts: "The American Philosophes did not have to be everlastingly on guard against the censorship of the State, the Church, and the University; they did not have to take refuge in evasion or suterfuge.... They could write boldly; they could attack the King, the Church, property—whatever they would—without fear of reprisal."[3][2]

As long as the librarians of America, individually, as a profession, as a force in American life, agree to follow actively—not just with lip service—this basic tenet of the American creed, then libraries—whether computerized, miniaturized, televised, or however technologically altered—will be a mighty fortress for the protection of the freedom of the human mind, for the support of those who wish to go wherever that freedom allows them.

# NOTES

[1] Described in Hideaki Okamoto, "Technological Change and Social Order," in *The Social Implications of the Scientific and Technological Revolution* (Geneva: Unesco, 1981), pp. 326-40.

[2] Dan Lacy, *Books and the Future: A Speculation* (New York: New York Public Library, 1956), p. 8.

[3] Lacy, p. 10.

[4] Ibid., p. 27.

[5] Ibid.

[6] Rule, p. 201.

[7] Ganley, p. 29.

[8] Ibid., p. 31.

[9] Ibid., p. 34.

[10] N. Bruce Hannay and Robert E. McGinn, "The Anatomy of Modern Technology: Prolegomenon to an Improved Public Policy for the Social Management of Technology," *Daedalus* (Winter 1980): 25-53.

[11] Ibid., pp. 32-34.

[12] Ibid., p. 34.

[13] Robert Louis Stevenson, "On the Choice of a Profession," in *Lay Morals and Other Essays* (New York: Charles Scribner's Sons, 1923), p. 257.

[14] Lester Asheim, "Ortega Revisited," *Library Quarterly* (July 1982): 217.

[15] Ibid., p. 219.

[16] Jose Ortega y Gasset, *The Mission of the Librarian* (Boston: G. K. Hall, 1961), p. 19.

[17] Asheim, p. 220.

[18] Ibid., p. 221.

[19] Ibid., p. 222.

[20] Ibid., p. 223.

[21] Ibid., p. 224.

[22] Eric Hoffer, *Between the Devil and the Dragon* (New York: Harper & Row, 1982), p. 12.

[23] Colin Norman, *The God That Limps* (New York: W. W. Norton, 1981), p. 78.

[24] Ibid., p. 86.

[25] Ibid., p. 140; citing Robert Howard, "Brave New Marketplace," *Working Papers* (Nov./Dec. 1980).

[26] Rudolph Bold, "Librarian Burn-Out," *Library Journal* (Nov. 1, 1982): 2048-51.

[27] Ibid., p. 2049.

[28] Norman, p. 112.

[29] Asheim, p. 220.

[30] Eli M. Oboler, "Attitudes on Segregation; How ALA Compares with Other Professional Associations," *Library Journal* (Dec. 15, 1961): 4233-39.

[31] Charles H. Brown, "College Library Objectives and Their Attainment," in *College and University Librarian and Librarianship* (Chicago: American Library Association, 1946), p. ix.

[32] Henry Steele Commager, *The Empire of Reason* (New York: Anchor Press/ Doubleday, 1978), pp. 265-66.

# BIBLIOGRAPHY

What follows is an alphabetical listing of the items indicated in this book's footnotes, but it is by no means a complete list of the items in the ever-growing bibliography on this important topic.

Abbott, R. P., et al. *Security Analysis and Enhancements of Computer Operating Systems.* Washington, DC: National Bureau of Standards, 1976. Report No. NBSIR-76-1041.

Agar, Herbert. *The Perils of Democracy.* Chester Springs, PA: Dufour Editions, 1965.

Allen, Charles. "Talking with a Soviet Free Trade Unionist." *New Leader* (Sept. 21, 1981): 6-9.

Arthur D. Little, Inc. *Into the Information Age: A Perspective for Federal Action.* Chicago: American Library Association, 1978.

Asheim, Lester. "Ortega Revisited." *Library Quarterly* (July 1982): 215-26.

Baker, Seth H. "Foretaken." *Data Processor* (June/July 1980): 10-12.

Barabla, Vincent P. "The Future Role of Information in American Life." In Norman Cousins, ed. *Reflections of America: Commemorating the Statistical Abstract Centennial.* Washington, DC: U.S. Bureau of the Census, 1980. pp. 193-202.

Beard, Robin. "Unesco Licensing." *Congressional Record* (Sept. 17, 1981): H6348.

Becker, Joseph, ed. *Proceedings of the Conference on Interlibrary Communications and Information Networks.* Chicago: American Library Association, 1971.

Bell, Daniel, ed. *Toward the Year 2000: Work in Progress.* Boston: Beacon Press, 1969.

Blake, Fay, and Jan Irly. "The Selling of the Public Library." *Drexel Library Quarterly* (Jan.-April 1976): 149-58.

Bold, Rudolph. "Librarian Burn-Out." *Library Journal* (Nov. 1, 1982): 2048-51.

Brier, Alan, and Ian Robinson. *Computers and the Social Sciences.* New York: Columbia University Press, 1974.

Brown, Charles H. "College Library Objectives and Their Attainment." In *College and University Libraries and Librarianship.* Chicago: American Library Association, 1946.

Bury, J. B. *The Idea of Progress.* New York: Dover Publications, 1955. [Originally published in 1920.]

Carnegie Commission on Higher Education. *The Fourth Revolution: Instructional Technology in Higher Education.* New York: McGraw-Hill, 1972.

Carnegie Council on Policy Studies in Higher Education. *Three Thousand Futures: The Next Twenty Years for Higher Education.* San Francisco: Jossey-Bass, 1980.

Collinson, Robert. "Trends Abroad: Western Europe." *Library Trends* (July 1970): 115-21.

Commager, Henry Steele. *The Empire of Reason.* New York: Anchor Press/Doubleday, 1978.

Council on Library Resources. *Linking the Bibliographic Utilities: Benefits and Costs.* Washington, DC: The Council, 1980.

DeGennaro, Richard. "Copyright, Resource Sharing, and Hard Times: A View from the Field." *American Libraries* (Sept. 1977): 430-35.

DeGennaro, Richard. "From Monopoly to Competition: The Changing Library Network Scene." *Library Journal* (June 1, 1979): 1215-17.

DeGennaro, Richard. "Libraries, Technology, and the Information Marketplace." *Library Journal* (June 1, 1982): 1045-54.

Dertouzos, Michael L., and Joel Moses, eds. *The Computer Age: A Twenty-Year View.* Cambridge, MA: MIT Press, 1979.

Dorn, Philip H., "Programs Are Not Books." *Datamation* (Nov. 1977): 231-33.

Dranov, Paula. *Microfilm: The Librarians' View, 1976-77.* White Plains, NY: Knowledge Industry Publications, 1976.

Durant, Will. *The Reformation.* New York: Simon and Schuster, 1957.

"Education and the Telefuture." *Change* (Nov.-Dec. 1979): 12-13.

Eisenstein, Elizabeth. *The Printing Press as an Agent of Change.* New York: Cambridge University Press, 1979. v. 1 and 2.

Fenwick, Millicent. "The Declaration of Talloires." *Congressional Record* (Sept. 17, 1981): H6354-56.

"For Universities, It's Harder Times in a More Complex World." *New York Times* (June 7, 1981): 21EY.

Fried, Rivka. "Begin Squeezes Israeli TV." *New Statesman* (Nov. 21, 1980): 12.

Ganley, Oswald H., and Gladys D. Ganley. *To Inform or to Control? The New Communications Networks.* New York: McGraw-Hill, 1982.

Garvey, William. *Communication: The Essence of Science.* New York: Pergamon Press, 1979.

Gastil, Raymond D., with Leonard R. Sussman. *Freedom in the World: Political Rights and Civil Liberties 1980.* New York: Freedom House, 1981.

Hannay, N. Bruce, and Robert E. McGinn. "The Anatomy of Modern Technology: Prolegomenon to an Improved Public Policy for the Social Management of Technology." *Daedalus* (Winter 1980): 25-53.

Harney, Joan M. *Specialized Information Centres.* Hamden, CT: Linnet Books, 1976.

Hartmann, G. K. *The Information Explosion and Its Consequences for Data Acquisition, Documentation, and Processing.* Washington, DC: National Academy of Sciences, World Data Center A, 1978.

Hecksher, August. "The Invasion of Privacy (2): The Reshaping of Privacy." *American Scholar* (Winter 1958-59): 11-20.

Hicks, Granville. "The Invasion of Privacy (3): The Limits of Privacy." *American Scholar* (Spring 1959): 185-93.

Hoffer, Eric. *Between the Devil and the Dragon.* New York: Harper & Row, 1982.

Holton, Gerard, and Robert Morison, eds. "Limits of Scientific Inquiry." *Daedalus* (Spring 1978): 1-236.

"How Silicon Spies Get Away with Copying." *Business Week* (April 21, 1980): 178-88.

Howard, Robert. "Brave New Workplace." *Working Papers* (Nov./Dec. 1980): 21-31.

Hutchins, Robert Maynard. *The Higher Learning in America*. New Haven: Yale University Press, 1936.

Johnson, Gerald W. "The Invasion of Privacy (4): Laus Contemptionis." *American Scholar* (Autumn 1959): 447-57.

Lacy, Dan. *Books and the Future: A Speculation*. New York: New York Public Library, 1956.

Lancaster, F. Wilfrid, ed. *The Role of the Library in an Electronic Society*. Urbana-Champaign: University of Illinois, Graduate School of Library Science, 1980.

Lewis, Paul. "Gloves Come Off in Struggle with Unesco." *New York Times* (May 24, 1981): 3E.

Malincomico, S. Michael, and Paul J. Fasana. *The Future of the Catalog: The Library's Choices*. White Plains, NY: Knowledge Industry Publications, 1979.

Marke, Julius M. *Copyright and Intellectual Property*. New York: Fund for the Advancement of Education, 1967.

Martin, James. *Telematic Society: A Challenge for Tomorrow*. Washington, DC: U.S. News and World Report, 1971.

McKeon, Richard. "Censorship." In *Encyclopaedia Britannica*. Chicago: EB, Inc., 1974. 15th edition. v. 3: 1083-90.

McLuhan, Marshall. *Understanding Media: The Extensions of Man*. New York: New American Library, 1964.

Miles, Bill. "From Prostitutes to Meter Maids—Unholy Sources of Urban Information." *RQ* (Fall 1978): 13-18.

Montague, Eleanor. "Automation and the Library Administrator." *Journal of Library Automation* (Dec. 1978): 313-23.

Montgomery, Edward B., ed. *The Foundations of Access to Knowledge: A Symposium*. Syracuse, NY: Division of Summer Services, Syracuse University, 1960.

"More U.S. Funds Urged for Basic Research." *Higher Education and National Affairs* (Jan. 11, 1980): 3.

Morphet, Edgar L., and Charles O. Ryan. *Implications for Education of Prospective Changes in Society*. Denver: Designing Education for the Future Project, 1967.

Mumford, Lewis. *The Myth of the Machine: Technics and Human Development.* New York: Harcourt, Brace, 1967.

Mumford, Lewis. *The Myth of the Machine: The Pentagon of Power.* New York: Harcourt, Brace, Jovanovich, 1970.

"NBS Study Recommends Copyright Changes to Protect Computer-Readable Works." *Journal of Library Automation* (March 1978): 74.

Newman, Joseph, ed. *Wiring the World: The Explosion in Communications.* Washington, DC: U.S. News & World Report Books, 1971.

Norman, Colin. *The God That Limps.* New York: W. W. Norton, 1981.

Oboler, Eli M. "Academe, The Library, and *Accidie.*" *Catholic Library World* (March 1982): 344-46.

Oboler, Eli M. "Attitudes on Segregation: How ALA Compares with Other Professional Associations." *Library Journal* (Dec. 15, 1961): 4233-39.

Oboler, Eli M. *Defending Intellectual Freedom: The Library and the Censor.* Westport, CT: Greenwood Press, 1980.

Oboler, Eli M. "From Books to Bytes." *Library Journal* (Oct. 1, 1981): 1877-80.

Oboler, Eli M. "International Aspects of Intellectual Freedom." *Drexel Library Quarterly* (Winter 1982): 95-100.

Oboler, Eli M. "The Challenge to Library Reference Service in the Decades Ahead." *The Reference Librarian* (Fall/Winter 1981): 55-58.

Oboler, Eli M. *The Fear of the Word: Censorship and Sex.* Metuchen, NJ: Scarecrow Press, 1974.

Oboler, Eli M. "Too Big for Their Britches: Or, How Useful Is a Utility?" *Technicalities* (Jan. 1981): 15-16.

Oceana Editorial Board. *Privacy—Its Legal Protection.* Dobbs Ferry, NY: Oceana Publications, 1976. Revised edition.

Oettinger, Anthony G. *Run, Computer, Run: The Mythology of Educational Innovation.* Cambridge, MA: Harvard University Press, 1969.

Ortega y Gasset, Jose. *The Mission of the Librarian.* Boston: G. K. Hall, 1961.

*The Poems of George Meredith.* Phyllis B. Bartlett, ed. v. 1. New Haven: Yale University Press, 1978.

Rees, Alan M., "Librarians and Information Centers." *College and Research Libraries* (May 1964): 200-204.

Righter, Rosemary. *Whose News? Politics, the Press, and the Third World.* New York: Times Books, 1978.

Rovere, Richard H. "The Invasion of Privacy (1): Technology and the Claims of Community." *American Scholar* (Autumn 1958): 413-21.

Rule, James B. "The Future of Freedom: Politics and Technology." *Dissent* (Spring 1982): 201-6.

Saltman, Roy G. *Computer Science & Technology: Copyright in Computer-Readable Works: Policy Impacts of Technological Change.* U.S. Department of Commerce, National Bureau of Standards. NBS Special Publication, Washington, DC: GPO, 1977.

Sarton, George. *The Appreciation of Ancient and Medieval Science during the Renaissance (1450-1600).* Philadelphia: University of Pennsylvania Press, 1955.

Saudek, Robert. "TV's Future Is Reflected in Print's History." *New York Times* Aug. 24, 1980): 33-34D.

Schmemann, Serge. "Commissars of Culture Don't Relax Very Often." *New York Times* (Nov. 8, 1981): 8E.

Schramm, Wilbur, ed. *Mass Communications.* Urbana: University of Illinois Press, 1960.

Siebert, Frederick Seaton. *Freedom of the Press in England 1476-1776: The Rise and Decline of Government Control.* Urbana: University of Illinois Press, 1965.

Simpson, G. S., Jr. "Scientific Information Centers in the United States." *American Documentation* (Jan. 1962): 43-48.

Smith, Ralph Lee. *The Wired Nation; Cable TV: The Electronic Communications Highway.* New York: Harper & Row, 1972.

Spero, Joan Edelman. "Information: The Policy Void." *Foreign Policy* (Fall 1982): 139-56.

Starr, Paul. "Transforming the Libraries: From Paper to Microfiche." *Change* (Nov. 1974): 39.

Steinberg, S. H. *Five Hundred Years of Printing.* 2nd ed. Baltimore: Penguin Books, 1966.

Stevenson, Robert Louis. *Lay Morals and Other Essays*. New York: Charles Scribner's Sons, 1923.

Thomas, Donald. *A Long Time Burning: The History of Literary Censorship in England*. New York: Frederick A. Praeger, 1969.

"TV to Homevideo, 1980-1985." *Variety* (Sept. 24, 1980): 49.

"Two Celebrations of Free Speech." *New York Times* (June 11, 1978): 6.

U.S. Congress. Office of Technology Assessment. *Computer-Based National Information Systems: Technology and Public Power*. Washington, DC: GPO, 1981.

U.S. Dept. of Commerce. National Technical and Information Service. *Telecommunications & Society, 1976-1991*. Springfield, VA: NTIS, 1980 [dated June 22, 1976].

U.S. General Accounting Office. *Challenges of Protecting Personal Information in an Expanding Federal Computer Network Environment*. Washington, DC: General Accounting Office, 1978. GAO Report to Congress 76-102.

U.S. House of Representatives. Committee on Science and Astronautics. *The Management of Information and Knowledge*. Washington, DC: GPO, 1970.

U.S. Library of Congress. Network Development Office. *Document Delivery*. Washington, DC: Library of Congress, Network Development Office, 1982.

U.S. National Commission on Libraries and Information Science. *Toward a National Program for Library and Information Services: Goals for Action*. Washington, DC: GPO, 1975.

U.S. President's Science Advisory Committee. *Science, Government, and Information*. Washington, DC: GPO, 1963.

Unesco. *The Social Implications of the Scientific and Technological Revolution*. Geneva: Unesco, 1981.

Veaner, Allen B. "Microfilm and the Library: A Retrospective." *Drexel Library Quarterly* (Oct. 1975): 1-5.

Volney, Count de. *Les Ruines des Empires*. Paris, 1789.

Voudran, Raymond V. *National Union Catalog Experience: Implications for Network Planning*. Washington, DC: Library of Congress, 1982. Network Planning Paper No. 6, 1980.

Westin, Alan F. "Life, Liberty, and the Pursuit of Privacy." In Donald P. Lauda and Robert D. Ryan. *Advancing Technology: Its Impact on Society*. New York: Wm. C. Brown Co., 1971. pp. 360-75.

Whitney, Frederick C. *Mass Media and Mass Communications Society*. Dubuque, IA: Wm. C. Brown, 1975.

Wicklein, John. "The Long Shadow of Censorship." *Atlantic Monthly* (Aug. 1979): 13-21.

Wilcox, Alice. "Resource Sharing." In Robert Wedgeworth, ed. *ALA World Encyclopedia of Library and Information Services*. Chicago: American Library Association, 1980. pp. 479-82.

Williams, Mitsuko, and Elizabeth R. Davis. "Evaluation of PLATO Library Instructional Lessons." *Journal of Academic Librarianship* (March 1979): 14-19.

Winner, Langdon. *Autonomous Technology: Technics—Out-of-Control as a Theme in Political Thought*. Cambridge, MA: MIT Press, 1977.

Winner, Langdon. "Do Artifacts Have Politics?" *Daedalus* (Winter 1980): 121-36.

Zimmerman, R. G. J. "A Computer-Accessed Microfiche Library." *Journal of Library Automation* (Dec. 1974): 290-306.

# INDEX

3